Our Tactile Brain Computed World and
The Platonic Brain Web Wikipedia

Our Tactile Brain Computed World and The Platonic Brain Web Wikipedia

PERCEPTION EXPERIENCES ARE DISPLAYS OF THE BRAIN'S ONGOING COMPUTATIONS, PROMOTED BY ELECTROCHEMICAL STIMULI

● ● ●

J. N. Schad, PhD

A DISCOURSE ON PERCEPTIONS

ISBN-13: 9781533451484
ISBN-10: 1533451486
Library of Congress Control Number: 2016908822
CreateSpace Independent Publishing Platform
North Charleston, South Carolina

Table of Contents

Ideas are imaginations produced by sensations or memory; thought is sequence of such imaginations. That sequence is not determined by free will but by mechanical laws governing the association of ideas.

THOMAS HOBBES (1588–1679)

Everything important has been said before.

ALFRED NORTH WHITEHEAD (1861–1947)

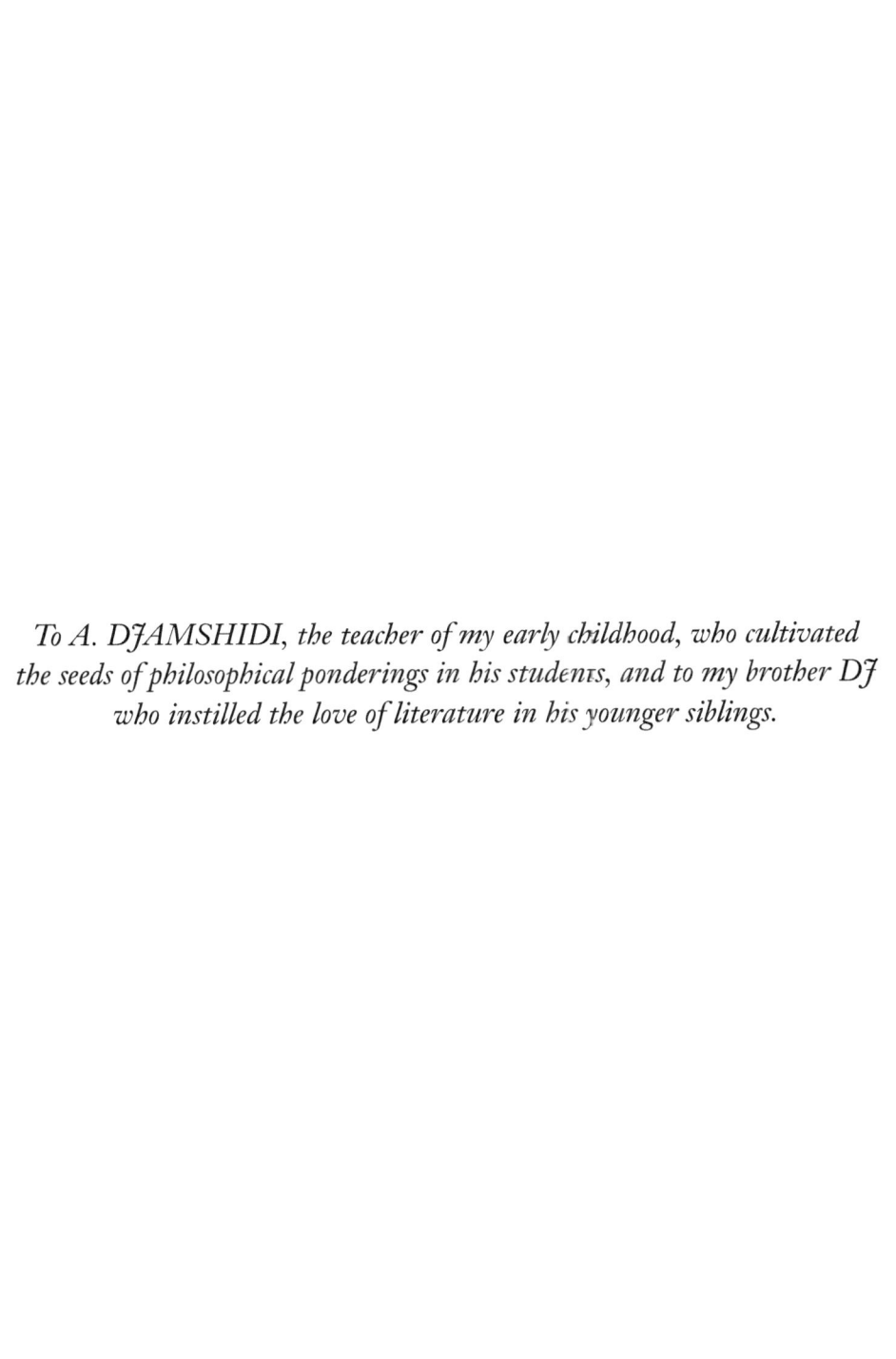

To A. DJAMSHIDI, the teacher of my early childhood, who cultivated the seeds of philosophical ponderings in his students, and to my brother DJ who instilled the love of literature in his younger siblings.

Acknowledgment

● ● ●

THIS BOOK IS CREDITED TO ALL DEEP THINKERS AND SEERS IN HUMAN HISTORY.

The Author

● ● ●

JAHAN N. SCHAD IS A PhD graduate of UC Berkeley in engineering sciences. His professional life encompasses works in scientific research, teaching, and industry. In 1998, after early retirement from almost twenty years of mathematical modeling of various engineering and applied physics phenomena at LBNL (UC Berkeley), he started work in the memory industry in Silicon Valley. Here, up until 2005, he helped manage the engineering quality of electronic chips for hard disks manufactured by Quantum and Maxtor companies. Presently he is engaged in quality electronic engineering in a consulting role. Along with engineering and scientific interests, he has pursued a lifelong fascination in mythology, mysticism (pseudoscience aside), religion, philosophy, and history, along with neurosciences and neural and neuronal networks computation, all geared toward gaining a better understanding of the workings of the brain. His decades of engineering experience and pioneering work resulted in published scientific research aimed at formulating and solving complex physical phenomena, and his lifetime curiosity and endeavors in learning and studies—partly reflected in some sixty peer-reviewed publications—have provided him the background and vision needed to take on the challenge of publishing this very unique work.

It is a major step in addressing some of the hard questions of life based on his published fundamental discovery of the underlying essences of the computational operation of the brain's neuronal network.

Preface

• • •

We feel clearly that we are only now beginning to acquire reliable material for welding together the sum total of all that is known into a whole; but, on the other hand, it has become next to impossible for a single mind fully to command more than a small specialized portion of it. I can see no other escape from this dilemma (lest our true who aim be lost for ever)than that some of us should venture to embark on a synthesis of facts and theories, albeit second- hand and incomplete knowledge of some of them—and at the risk of making fools of ourselves.

PHYSICIST ERWIN SCHRODINGER

IT IS NOT LIKELY THAT we will ever convincingly know how and why we came to be on this planet; of course, this has never prevented inquisitive minds from pushing the frontiers of understanding and discovery further. Our origin is the subject of scientific theories and continuous inquiries with no end in sight, as the shells of related complexities are getting much harder to crack. Paraphrasing philosopher and historian Will Durant, a very few people are getting to know more and more about less and less. This knowledge and the scientific language used to express it are now becoming more and more incomprehensible to most humans; therefore, despite the sincere efforts of communication media's countless "talking heads," we may be inevitably driven by default to a new religion of worshiping the church of science, along with its scientist "priests," and to a life filled with the unequivocal dangers that could befall us from such devotions. Humanity and the planet are faced with dangers that have and can be issued from the absence or deficiency of proper public knowledge and guiding ethical principles for handling many of the new scientific and engineering discoveries and innovations, such as artificial intelligence, nanotechnology, and plant, animal, and human gene editing, etc. To avoid inadvertent adverse impacts, which could seriously endanger our normal ways of life and even existence, knowledge has to be humanized and massively disseminated. Even at the expense of some mistakes, the benefits certainly outweigh the loss.

This book is a small effort along this path.

Introduction

● ● ●

Let us enter within, if we can fair out the
ultimate nature of our mind, we shall perhaps
have the key to the external world.

A<small>RTHUR</small> S<small>CHOPENHAUER</small>

T<small>HOUGH MOST HUMANS HAVE LIVED</small> believing that their lives are what gods/God/destiny/nature intended for them, their innate curiosity, perhaps a device of the evolution schema, has resulted in discovering some of the workings of their universe, as well as the workings of their own beings. These discoveries have come about as the result of their deep introspections and mental (brain) efforts to understand the natural phenomena—those affecting them or affected by them. Our civilization owes its present status to such collective mental endeavors, though there is much to be desired. On the face of it, this gained knowledge, from the children of gifted, curious brains—from Zarathustra (Avey, 1954) to pre- and post-Socratic philosophers, and scientists and humanists of all eras—should have seemingly settled and guaranteed a peaceful existence for humans and a harmonious coexistence with nature, nondestructive and conservative! This end has long been within reach, given the availability of brilliant life philosophies, ideologies, social theories, and technologies, which could shortcut the arduous path of progress for all humanity. However, they have not so far been seriously and effectively heeded, and the goal still remains a far-reaching dream, despite the untiring efforts of many enlightened humans, which continue to present day. Unfortunately, the problems of humanity persist; future perils of climate change, population increases, and food shortages, etc., are lurking. The momentary glitters of life in modern societies have blinded mankind from the long-term legitimate concerns for future sustenance, partly due to conflicts with the habitual ways of life of many, and mainly because of the pernicious collective intentions of some who have sought satisfaction in their boundless avarice, regardless of the havoc it wreaks upon others and the irrecoverable costs and uncorrectable damages to the environment.

Although the main underlying reason for the prodigious obstacles on the path of compassionate sustainability appears to be the lure of wealth and power with Machiavellian mind-sets behind them on the one hand and the ordinary sociopolitical illiteracy of the masses on the other, nonetheless, ignorance—the lack of true knowledge, at all levels, even in the case of society's most powerful—is the force that renders insensitivity to the fact that eventually all of humanity, and our posterity, will be suffering from the consequences of a short-sighted or misguided (individual or collective) philosophy of life.

Of course, there must be deeper explanations and rationales from anthropological, as well as ontological, viewpoints. It is a fact—the ultimate justification—that humans are just in the dawn of their civilization, which implies that they are in their very early mental growth stages as well, given the slow pace of the evolutionary processes. Therefore, it may take many millennia before natural and societal selection, hand in hand, render the needed balance between the frontiers of knowledge and what it precipitates in the way of behavior of the masses, aligned with their collective and individual conscious wants and the actions they deem justifiable for attaining what they consider better lives, all part and parcel of the overall picture of human existence.

When, and if, all humanity is driven by the brightest minds' noble causes, the race to attain universal equanimity and harmony will be much accelerated. The resources of the brain and its treasure of knowledge are at man's disposal to serve him along his path, and perhaps shape him, at some point, in the image of the nobility that Rousseau was ascribing to the survival-driven savage, provided his early mental shortcomings do not destroy his very existence before the opportunity presents itself. The fact remains that the prospect of a disastrous end could be circumvented if the accelerated pace of knowledge dissemination—prevalent in the period lasting from the Enlightenment era to the early twentieth century, which has suffered setbacks in the recent past and is now following a slow, diffusive pace—regains its footing. Much speedier awareness is needed to compensate for the now-understood ill consequences of exploitive industrialization, a phenomenon that humanity did not much suffer from before its onset. Enlightenment at all levels of humanity is the key to the balanced and optimal continuation of existence.

What may make one aspire to such possibility are today's achievements in the fields of humanities and sciences in general and the understanding of the workings of the brain—as limited as it is—in particular, which may help clarify what underlies human collective behavior in its widest connotations, what has brought humanity to its present status, and how this can be affected for the desired end. Seemingly, the powers of innate primal and acquired knowledge, and their skeweness, must have been behind it all. These forces in Schopenhauer's philosophy are the ***unconscious will*** and intellect, and their role defined metaphorically is the ***"will is the strong***

blind man who carries on his shoulders the lame man who can see. " Whether we agree with him or not, the intellect, the acquired knowledge, is the guiding light. Divulging of the brain mechanisms and the processes of acquiring, imbedding, and using knowledge— clarification as to the natures and mechanisms of perceptions and subjective experiences, such as consciousness, thought, free will, etc. (the big questions of philosophy, the elements that define life and beings, or as Schopenhauer puts it, ***"the world idea"***)—are steps toward enlightenment, which we embark on helping.

In this book all above elements will be addressed in the context of functional computational operations of the brain—its complexity resolution and solution schemes (Schad 2016)—and in the process, light will be thrown on some puzzling secrets and dilemmas of life. The work will provide an analytical view of man's mental powers in the context of the machinery of the brain, which drives it. Attempts will be made to develop a realistic perspective on who we are and what our potential is. In light of the premise just stated, various aspects of the mind will be addressed in order to drive home a beautiful ***"theory of simulating brain"*** and the concept of ***"the inner platonic world, the Brain Web Wiki."*** A deep understanding of these concepts may provide truth seekers with the answers they yearn for.

Finally, I think the most important contribution of this book will be felt when the reader experiences his or her own inner "Aha!" moment about life. I hope to have made a strong case that the answers to life's most difficult questions may perhaps be found in the back alleys of the brain, as long as the questions are correctly formulated and posed.

The Brain Web Wiki is the extreme encyclopedia of life!

Background

• • •

Omnibus ex nihil ducendis sufficit unum.

Physicist John A. Wheeler

The key characteristic of the conscious mind, or personalized brain, is its persistent engagement with why, where, when, what, and how, in relation to life and the elements affecting it or affected by it. The depth to which such questions are posed depends on the level or maturity of consciousness and on the mind's measure of disengagement from the pressures of survival and the material aspects of living and life. Such inquiries must have begun with early human's awe of nature and dramatic natural events; while instilling fear and respect in them, these factors also helped to improve the brains over the course of multitude generations. Human's growing curiosity, concentration, and interests planted the seeds of what, in the course of several millennia, became mythology, religion, philosophy, and their eventual offshoots—the sciences.

The mythologies and the emergence and evolution of religions among different cultures of the world—being mostly amazingly similar at their depth (Joseph Campbell) and unfathomable by rational thinking—are thought to be the results of experiences of altered

1

states of awareness caused by certain approaches to seeking the truth, such as long and sustained meditation, the use of hallucinogenic plants, and so on, which temporarily disengage the brain from the normal concerns and constraints of living. The rich writings of ancient times, such as the Vedic hymns of the early Aryans, which address the creations, revelations, and reflections of ancient men on the mysteries of nature, may possibly relate to such states of mind. Concepts such as biological evolution on Earth, energy as the building block of existence, hints of the existence of nuclear energy, the notion of cosmological time, and many others all derive from the surreal discoveries of some of the earliest wonderers, often known as mystics. (They were also generally known for the other creative facets of their lives.) Persisting questions about life, the creator and creation, souls, minds, and matter, and of humanity's depictions of them, which affected beliefs, behavior, formation, and the governance of societies, were on the minds of the deep thinkers, who dedicated their lives to addressing them. The answers they arrived at formed the basis of their philosophies.

Zarathustra's philosophy of life, which set the stage for the works of pre-Socratic thinkers and was later taken up by Socrates, Plato, and Aristotle, laid the groundwork for the flowering and evolution of philosophical thought up until the present time. For almost a thousand years, from the fall of the Roman Empire to the Renaissance, Aristotle's work served as the basis for most of the physical and metaphysical inquiries of thinking individuals, made largely in the service of religion. The sixteenth century marked the beginning of a new era of post-Aristotelian philosophy. In this period, all prior philosophical thoughts and beliefs were reexamined. This process created mental challenges for the thinking man. The last few centuries' progression of philosophical works and their profound effects on various philosophical views are explained beautifully by a statement of

Will Durant in his "The Story of Philosophy," which is condensed and paraphrased in the following paragraph:

The key to the evolution of philosophical thought was structured logic, with its solid forging of ethics, esthetics, politics, and metaphysics, which began with Socrates, among others, and was fully realized by Aristotle. Aristotle's logic and philosophy regarding nature and the nature of things lasted for almost one thousand years, shaping medieval scholarship and culminating in the works of Thomas Aquinas (1225–1274 AD). Aquinas's influence continued throughout the Renaissance (1400–1600 AD) and remained strong until the nineteenth century, and all aspects of his philosophy are studied up to the present day. The Renaissance period started with the rediscovery of classic Greek philosophy through translations of Persian (Arabic version) and Jewish texts; this led to breaking out of the shell of the religiously transmogrified Aristotelian philosophy and to new thinking, which manifested itself in arts and innovations that laid the basis for the advent of reason. Beginning with the sixteenth century, commonly referred to as the modern era, philosophers of "pure reason," prominently Bacon and Voltaire, did away with most of what the philosophers of Judaism, Christianity, and Islam had hanged their metaphysical hats on and decided that all phenomena could be explained by the sciences. But the deep roots of religion and what was built around it, including fear of loss of faith, soon forced a reexamination of pure reason. Efforts of rationalists and empiricists such as Locke, Berkeley, and Hume, though still not so favorable to religion, left pure reason in its own ruin, while the material and nonmaterial soul were also slain by its sword. This ravage of reason also sacrificed the laws of science, since events and sequences precipitate into perception, which is changeable. Everything, with the exception of true and unchangeable mathematical formulae, was just a set of ideas stemming from sensory experience. Science and faith, left to the skeptics, were both about to be sacrificed at the altar of experience. Meanwhile, the philosopher Rousseau almost single-handedly questioned logic as a new, frail, and deceptive part of us, one that could not be trusted. According to Rousseau, not much should be left to reason that fails to take into account innate human emotions and feelings. In his rebellion against logic and reason, Rousseau claimed that culture is a corrupting power, "thinking man is a depraved animal," and intellect serves mostly to suppress morality. Instinct and feeling are more trustworthy than reason. This approach led to the Romantic literature of Europe, which was dominant for almost a century, as well as the revival of religion. In the philosophical chaos of arguments for or against either reason or instinct, Kant appeared to save religion from reason and science from skeptics. His book, "The Critique of Reason," almost miraculously put knowledge, mind, and matter in an elegantly consistent context, and this approach continues to reverberate even to the present time. Kant offered a tangible platform of thought, as a result of which science and faith have continued to enrich human lives.

With the awakening and enlightenment that started with Roger Bacon (1214–1294) and da Vinci (1452–1519), and continued through the works of great men of philosophy and knowledge such as Copernicus (1473–1543), Bruno (1548–1600), Galileo (1564–1642), Francis Bacon (1561–1626), Descartes (1596–1650), Hobbes (1588–1679), Spinoza

(1632–1677), Locke (1532–1674), Newton (1642–1726), Berkeley (1685–1753), Leibniz (1646–1719), Hume (1711–1776), Voltaire (1694–1778), Rousseau (1712–1778), Diderot (1713–1784), Kant (1724–1804), Fichte (1762–1814), Hegel (1770–1831), Schopenhauer (1778–1860), Comte (1798–1857), Kierkegaard (1813–1855), Darwin (1809–1882), William James (1810–1942), Spenser (1820–1903), Marx (1818–1883), Wundt (1832–1920), Nietzsche (1844–1900), Cantor (1845–1918), Husserl (1859–1938), Freud (1859–1936), Jung (1875–1961), Russell (1872–1976), Schrodinger (1877–1961), Einstein (1879–1950), Heidegger (1889–1976), Wittgenstein (1889–1951), Carnap (1891–1970), Pauling (1901–1994), Gödel (1906–1978), Sartre (1905–1980), and many others in the sciences, humanity has made great strides in solving many mysteries of life, a process that has kept pace with different scientific disciplines and a great many educational and scientific organizations, mostly established within the last two centuries. Beginning in the nineteenth century, massive efforts in education and research, afforded by stronger economies, created an environment and systems that were amenable to the finely channeled path of knowledge acquisition and the processes of discovery. Systematic approaches for unlocking the mysteries of the universe have resulted in fascinating scientific and technological achievements and today attract the brunt of inquiries into these subjects.

Although all that has been achieved during the last two millennia has to be ascribed to systematic logical thinking, initially at individual levels and later within the context of disciplined institutional systematic approaches, there has always been an irrational and mysterious counterpart unconsciously deployed in the processes of discovery and scientific progress: on many occasions the flash appearance of a solution or discovery is marked by a baffled "Aha!" or "Eureka!" from the researcher (Penrose 1990). Eminent scientists, discoverers, inventors, and composers of recent centuries have provided accounts of such experiences throughout their careers, though they have rarely dwelled on the reason for them, other than assuming that proper background preparation would eventually bring results. Though correct on the face of it, this assumption falls short of explaining how a solution or discovery actually occurs. Being a sudden event in the mind, it must be an end product of some process. Psychologist Julian Jaynes explains it this way:

> *There are several stages to creative thoughts: first, a stage of preparation in which the problem is concisely worked over, then a period of incubation without any conscious concentration upon the problem, and then the illumination which is later justified with logic.*

There are also seemingly nonmysterious phenomena of the mind—personalized brain—that partly characterize humans in the realm of their consciousness: these are knowledge, wisdom, and intelligence. Knowledge or knowing consists of information plus what the mind has made of it in terms of the development of perceptions and conceptions (thought), using a combination of earlier developed and innate (a priori) concepts, as Kant and Schopenhauer diligently point out in the case of the latter. Knowledge can be very deep-focused and limited to certain areas, or it can be multifaceted, covering many areas. In the former case, it defines certain expertise, and in the latter case, it is distinguished as wisdom, and by no means are they mutually exclusive. Intelligence is generally identified by the evidence of potential for fast learning and problem-resolution abilities, which is born out of knowledge, innate or otherwise, and/ or wisdom. And obviously perception—that is, making sense out of sensing—is the starting point of it all. Of course, such characteristics are only part of the complex package that defines humans and that has long been a matter of philosophical and scientific pursuit. Such efforts have resulted in more in-depth and fundamental questions being posed and a great many answers being found in the fields of philosophy, epistemology, and in particular the related fields of cognitive sciences, which have provided a wealth of brilliant concepts regarding the nature of thought (perception and conception) and thinking, whether it is algorithmic (of logical roots) or whether it also draws upon certain nonalgorithmic capabilities (perhaps using an as-yet not understood physical-mathematical approach). The latter possibility, in some respects, is fueled by some of the seeming characteristics of consciousness mentioned earlier, such as intuition, common sense, occurrence of ideas, solution of mathematically non-provable problems by intuitive approaches (Gödel), and finally the reported "Aha!" phenomenon.

Problem solving, though seemingly ascribed to out-of-the-ordinary mental practices, is the daily and even momentary engagement and stuff of the mind, of which one remains mostly unaware. It is in essence the mental processes one invokes to formulate thoughts, to handle various complexities encountered in living, or to meet the requisites for the conduct of life. The underlying mechanism can be better understood in the context of the role of the mind's major functional components. The components are the conscious mind and its unconscious counterparts, which together have evolved along the evolutionary path of life. All that is beyond one's consciousness is attributed to the operations of the unconscious: it carries the main burden of the inner (autonomous and otherwise) sustenance of life, as well as almost all (if not all) aspects of its animation. Its role can be likened to that of the operational systems of the most complicated automatic machinery one can imagine. Reasonable inference from the automation concept, as it relies on firmware, hardware, data storage, operating system, and software, would imply that the unconscious must comparably encompass all the necessary infra-structure and instructions—software, to speak metaphorically—which operate through the biological machinery that comprises the computational system of the brain. The following quote from Nobel Laureate neuroscientist Eric Kandel (2013) affirms this concept:

> *This new science of mind is based on the principle that our mind and our brain are inseparable. The brain is a complex biological organ possessing immense computational capability: it constructs our sensory experience, regulates our thoughts and emotions, and controls our actions. It is responsible not only for relatively simple motor behaviors like running and eating, but also for complex acts that we consider quintessentially human, like thinking, speaking and creating works of art. Looked at from this perspective, our mind is a set of operations carried out by our brain. The same principle of unity applies to mental disorders.*

The conscious component is the result (output) of the unconscious (computational) processing of the brain-imbedded instructions, ranging from those needed for survival and learned skills to those used for various protocols, rules, and information. And all these are acquired in response to the dictates of external, generational (evo-lutional), or lifetime environmental stimulations. Learning ability, sustained over the course of individual lifetimes and continued along the evolutionary time span, develops and augments the unconscious faculties in response to the living conditions encountered, to further supplement the natural selection process and enhance the survival prospects of the species.

A rather recent tool in the computer sciences field has the prospect of providing a mathematical basis for the findings of the neurosciences as to brain operations being computational: it is the development of the mathematical computational neural networks, patterned after the web of neurons in the brain, which provides the possibility for better understanding of the brain's computational operations and its power. Unlike digital computers, these synthetic neural nets demonstrate signs of what may loosely be called intelligence. They can be trained to solve problems without the apparent use of formal algorithmic instructions. A neural network behaves as a computational device that develops, remembers, and seemingly imbeds implicitly its own software. This capability can serve as a basis for envisioning similar properties and proofs for the neuronal net of the brain. Accordingly, the physical or geometrical configuration of the neurons, their plasticity, and the complex neuronal electrochemical phenomena instigated by various stimuli, which render various gene neurotransmitter expressions at their connections (synapses), would conceivably allow for the availability of all the elements needed for the computational and information-processing aspect of the brain.

Drawing a further parallel with the (brain-inspired scientific) mathematical neural network computers, the computational machinery of the brain can be fundamentally conceptualized here in such a likeness, called a neuronal net, though with a survival perfected/tuned physio-biological construct, which has evolved in the face of the complexities of nature that configured it (Kant was on the right track to suggest the presence of a structure inherent to it). The extensive power of the brain's information processing, in the face of vastly varying complexities and solution demands engendering both the conscious and unconscious aspects of the mind, must be based on the adjustment capability of the brain's neuronal web—that is, upon its ability to increase neuronal connectivity and vary (engaging more of and reconfiguring) the circuitry of neurons. This ability is called brain plasticity. It would accommodate for perceptions, learning ability (metaphorically, software imbedding), and network storage development for short- and long-term memory (the idea of a central memory in a specific part of the brain has long been abandoned). Based on such observations, very sophisticated computer-like (serial-parallel and perhaps quantum) operations reflect today's understanding of what is going on in the brain, which has been the subject of intensive research in various related fields.

The development of the neuronal web of the brain, the imbedding of appropriate (autonomic) computational controls (metaphorically, operation software and related knowledge), and the integration of peripheral systems are what mainly underlies the essence of the unconscious part of the brain. Since this process must take place in step with species evolution, it is inherently slow. However, the development of consciousness—that is, the growing awareness of the knowledge and skills necessary for living—is a rapidly evolving process that occurs over a species' lifetime. Therefore, it is thought to have little holdover potential for the immediately following generation. Nonetheless, the unconscious is very likely the depository of causative information about events and natural phenomena, controlled by natural laws (discovered and undiscovered), and developed as the results of life exposures of generations along the evolutionary time span.

In light of the given potential of the brain's computational facility, philosophy and the sciences must have had their roots in a measure of inherent knowledge among the species Homo sapiens. To such possibility, philosophers from Immanuel Kant to philosopher-scientist Albert Einstein have stated that the "mind is not a blank slate (tabula rasa) upon which experience writes." This was how Kant started to tackle the British school of Locke, Berkeley, and Hume, with its motto of "no Matter, no Mind." Kant singled out mathematics as the brain's invariant (unchangeable) product, which, since it has nothing to do with sense experience, provides proof of the inherent nature and structure of the brain. But I think the evolutionary implication of Hume's philosophy of the "mind being ideas in procession and association" is also behind some of the knowledge engrained in the brain. Such knowledge is the product of trial-and-error learning about the kind of perceptions that allow sustenance of life; it is the learning needed to correctly depict the life environment, to which the species' very survival attests. This learning occurred out of necessity, and what was learned through experiences in the evolutionary path was hacked (genetically expressed) into the developing brain as an unchangeable part (an actual brain structure and innate property of the mind); perhaps it is the mathematical essence of such neuronal configurations (Schad 2016) that also puts them in the realm of Kant's claim.

To further the computational arguments, consider the phenomenon of seeing as an example. It involves the capture of photons followed by a set of immensely complicated electrical-chemical processes that result in a proper image. This discernment and perception phenomenon here implies the presence of certain fundamentals, *imbedded* physical-mathematical principles required for the imaging process in the web of the brain. It is important to note that unlike many events in which natural laws dictate outcomes—as in the case of the flow of water in rivers or the flow of electricity in a wire—what is involved here is the reproduction of an image derived from the visual data the brain receives. In scientific lingo, this requires that the mathematical formulation of the vision phenomenon must have already been configured implicitly into the neural web of the brain, enabling it to do what is really a computational simulation (replication of an object), resulting from solving a vision problem that is posed (albeit subjectively) in the brain, subsequent to receiving reflections (input data) from the objects.

One need not look very far to find a proper rationale for the claim that simulations, i.e., solutions to a mathematical expression of a problem, are what really happen in the brain. The visual simulations experienced in dreams and hallucinations, prompted only by latent thoughts due to lack of external stimuli, are a further indication of the reality of simulation operations being performed by the unconscious mind. Not all computations have visual effects or lend themselves to other perceptions either. In reality, nearly all unconscious simulations remain unknown to the conscious mind. The brain's ongoing information processing and needed simulations (computations) are responses to real-time problems posed by everyday living. Here it should be emphasized again that the term "simulation" is used synonymously with the mathematical solution of problems governed by the physical laws, subject to certain domain (time-space) constraints, for specific circumstances, the results of some of which can also be the replication of objects and the environment as perceptions.

This emphasis on physics accords perfectly with the views of Reductionists, who believe that "all high-level governing sciences can be reduced to physics at the molecular level (expressed by mathematical language)." We may not, however, need to defend such views at all because realistic laws, at whatever level and of whatever nature, would be present implicitly in the computational web of the brain. As Kant puts it, *"Laws of thought are laws of things, for things are known to us through them that must obey these laws."* This concept is also implicitly confirmed by Hegel when he claims that *"Laws of nature and laws of logic are the same."*

Of course, at times the brain is taken to task, for example in cases like visualization on demand, hallucinations, dreams, or, last but not least, the processes of complex thinking. Such tasks are driven by various measures of internal or external stimuli that can force a result (solution). The degree of meaningfulness or accuracy of the simulations (solutions) appears to depend on how well the problem or question is posed. The relevancy of the solution, in fact, depends on posing (formulating) the right question. Obviously, thinking is at the heart of any question. Thinking involves intense activity within the brain, and systematic thinking like that required for disciplined education and research actually trains and preconditions the neural web by means of its own rules. Although such preconditioning is essential to the systematic solution of problems, its channeling effects (discipline dictate) on the brain may hinder freethinking and thus limit the use of the mind's full potential. The absence of this hindrance, allowing freethinking, may explain why earlier man could have (autonomously) experimented with the full potential of his brain in making the discoveries we earlier alluded to, albeit with a possibly less expanded and less configured unconscious machinery. At this juncture, it may be of some value to think that, regardless of the approach deployed in accessing brain's potential, *in the realm of the brain's computational design, mental formulations (scripts) of problems exist in high-level language of thought, which are then converted to the machine language of the brain for the solutions.*

All the greats in the field of psychology agree that, in one way or another, the mind is the storehouse of information and events. To some it is the library of memories—those related to the experiences of a single lifetime and/or those spanning the entire process of species evolution. The majority of this library's holdings, though seemingly not in the immediate reach of the consciousness, have proven to be accessible in parts, at least to some degree. The field of clinical psychology relies profoundly on the possibility of accessing the deeper psyche, the subconscious part of the data storehouse—recovering disassociated memories—which plays a role in shaping some conscious experiences. Much of the mind's current memory contents, especially those related to daily life events, are haphazardly patched together by deep-seated drives and emerge in the form of dreams.

Psychologist Carl Jung points to a much deeper part of the mind, the unconscious, as the holder of evolutionary secrets and mysteries (evolutionary memories). It is believed that some of the contents of the unconscious may be revealed as a result of the temporary disengagement of the brain with the external stimuli; in a mild case, this phenomenon would happen in a dreamlike state. During long-term disengagements, as in the schizophrenic state, one would live entirely, and sometimes permanently, in the world of these contents.

In a larger context, a willful disarming of the unconscious, cultivating absolute quiescence (uneventful consciousness) in order to allow a sharp focus, may be the key factor in unlocking some of the unconscious mind's contents, availing it to consciousness. Achievement of such a calming of the mind, except in channeled scientific research, is, however, beyond the reach of nearly all mankind. Today's accelerated pace of materialistic life is not making things any easier. Disengagement from the ordinary ways of life may require a lifetime effort. The noisy nature of the conscious mind, indicative of the heavy engagement of the unconscious machinery, is seemingly a tough barrier to the needed quiescence. Though the achievement of such a state of brain seems nearly prohibitive to modern humans, its exploits by ancient mystics—their psyches engraved only with the awe of nature—provide strong evidence of the possibility of overcoming the barrier and subsequently transcending the conscious mind. Heraclitus (sixth century BC) designates fire as the substance of the universe, what moves everything; the amazing poetry of the famed Persian mystic Rumi (tenth century) hints at the presence of immense power, similar to that of the sun, in the heart of every atomic nucleus. Such insight cannot be other than the achievement of such transcendences. At lower, yet more tangible levels, some dream experiences, being seemingly out of this world, occasionally happen, and they also could be attributed to transmissions from the deep unconscious to the conscious, triggered by chaotic neural bursts.

The diverse, extraordinary evidences of a nonpersonal (objective) nature in the unconscious mind point to an encyclopedic wellspring emanating from the genetic expressions of the tangles of the web of the brain. This is perhaps what Plato referred to as the world of knowledge, though in his perception, as great as it is, the world of knowledge remains an external one. The fact that the real-world laws of nature are the sculptor of the brain may, however, inferentially justify Plato's claim. From the scientific front, more piecemeal confirmation of stored instruction in the nervous system is turning up: it is now believed that motor development in beings results from an intertwining of ***information from genes,*** coded over the course of evolution, plus information from the natural environment. Many philosophical stances and ever-increasing hard scientific results are confirming the path of the ideas pursued here, so let's explore them in depth.

The Evolving Life

● ● ●

That which is in us pursues its ends
by the light of knowledge.

SCHOPENHAUER

A SPERM AND AN EGG meet (for mammals and metaphorically for most), and a commotion starts: many cells appear, each with a mission. Following their embedded instructions, an ensemble emerges. In final form, a machine, which in time will be capable of creating the very elements that created it, is developed. The cells, which are so produced and make up the final many forms of life, are little machines in themselves, which keep functioning until no purpose is served for doing so. That is when old age sets in and the potential for the onset of another generational round ends!

What is behind such an operation are the embedded cell scripts (blueprints), which are coded in many different combinations of four molecules, which are hung together variably in the double helix of DNA, known segments of which (about thirty thousand) are called genes, and the rest remains an enigma. Various transcriptions of genes favored by nature—entailing various functional scenarios, influenced by their chemical surroundings, and dependent on their varying missions evolved through natural selection—have rendered all the complexities and variations in the life-forms. The life blueprints (the DNA) of species pass to their subsequent generation through reproductive cells and remain mostly unchanged, though, subject to occasional gene mutations. And this may bring in a host of possibilities engendering changes in beings, ranging from more adaptability to their natural environment, to the species diversity that we have today—all at the mercy of an unknown game that we call chance.

To explore how it all started, we have to go back—way back—when a single-cell creature appeared, for one reason or another, upon the early Earth. It is now accepted that the needed chemicals for such a development existed in the primordial soup (hospitable chemical-rich watery bodies) in the early Earth environment, and, therefore, it need not have come from anywhere else, such as outer space, as some had speculated. Based on the scientific understanding of evolution today (Darwinism), the single cell gradually changed into a multicell creature, perhaps due to the tendencies of the emergent phenomenon born out of chaotic conditions, and started its journey on the evolutionary path to render the variety of species there are, as time marched away from such a humble beginning.

Considering the evolutionary path of life, time and time keeping would be embedded in the machinery of the cells, including those of the nervous system (when present), as it has now become scientifically determined (DNA Telomeres). As such, autonomous execution of the embedded instructions, is they the genetic codes or the evolutionary neural constructs (patterns), as a function of time result in changes in the living form and its functionality, preparing it for sustenance and survival in its environment, allowing it to cope with the phenomena affecting it, and ensuring the continuation of the species.

It is noteworthy to mention the fact that philosophers in the pre- and post-Aristotelian era understood the evolutionary process of nature, which must have laid the groundwork for the theory of evolution (Darwin 1809–1882). The beautiful poetry of Titus Lucretius (99–55 BCE) in his book *On the Nature of Things* speaks to such knowledge (quoted from Will Durant):

Many monsters too, the earth of the old tried to produce, things of strange face and limbs...

Some, without feet, some without hands, some without mouth, some without eyes...

Every other monster of this kind earth would produce, but in vain; for nature set a ban on their increase, they could not reach the coveted flower of age, nor find food, nor be united in marriage...

And many must have died out and been unable to beget and continue their breed, for in the case of all things which you see breathing the breath of life, either craft or courage or speed has from the beginning of its existence protected and preserved each particular race...

Those to whom nature has granted none of those qualities would be exposed as a prey and booty to others, until nature brought their kind to extinction.

The ensembles' bio-machines thus developed require energy for sustenance, in the least until the cycle of procreation transfers the role to the new generation. However, securing energy—where many machines of different varieties and capabilities compete for the very same resources—requires sophisticated abilities. And of course for all life-forms, whatever their stage of evolution, survival depends on their abilities commensurate with their biological complexities and needs thereof. There are tremendous variations in the functionalities of the faculties among beings. A main faculty that has fostered survival among many species is the communication aptitude that enables them to convey—mainly among their likes—their perceptions of their internal and external environments.

During the early evolutionary stages, hermitic life among creatures must have prevailed (it continues for some today), where each of the species separately struggled for survival, and some measure of learning had to take place, which would likely be registered (somewhere) in their systems. Putting it plainly, a "feel" for their environment—that is, their impressions of their encounters with nature and their experiences with the physical world, in terms of memories and skills—had to develop. And this allowed the creature to live isolated from others most of the time, except for the time of mating, for which few innate communication signals sufficed to attract the mate and perform the act. However, the commonality of the "feel," when grouping came about, had to be communicable in order to reinforce the very purpose of it, which is better survival. And this must have been the driving force behind the origination of some kind of language, at least in some species. However, communication, as primitive as it may have been, required a venue, a vehicle: touch alone, for many, would not meet the end, since separation for feeding would be inevitable. Therefore, detectable signals had to be issued to the external environment, using any possible mode, such as sound in the simplest form of vocalization, and it has evolved to the creature sounds we hear presently. In the case of some insects, chemical or motion signals turned out to also be a workable alternative.

The vocalization apparatus very likely has its roots in bodily vibration modes, which must have developed in the process of detection and feedback responses, resulting from sensing natural sound waves on the skin early in evolution. With the availability of vibration mechanisms, espoused by the activations of motor sensory neurons, the development of the biomechanical vocalization apparatus can be imagined. It must have happened as a result of certain mutations favoring the survival of a large branch of species. Later it evolved as an utterance system (vocal cords), along with the auditory system, continually engaged in the feedback and verification processes involving all senses, as well as the nervous system—whatever existed of it—in rendering proper representations of the environment in the brain.

In time, utterances began to carry meaning among groups, emerging as what is called by linguists as referential language. *The audible utterances, reflecting on their encounters with the environmental, that is, the receipt of and response to sense data, may have also, at some point, served the onset of the primitive loud thinking ability, as well.* Of course, inaudible utterances must have (much) predated audible ones. The immanent outputs of the brain's operations in the management of life—motor signals requiring the presence of a venue, which must have served as a predecessor to vocal cords—would dictate it.

The above presumptions are based on the belief, among philosophers and scientists, that many beings are capable of thinking, which is based on the neurophysiological potential of the brain along with observations. And further, that as a consequence of this ability, one way or another, these beings have developed a certain way (a language) to convey it, a language that relates mainly to the physical world. However, in more advanced ones, such as mammals, which have the benefit of the neocortex, it would also involve emotional aspects, which must be uttered in codes or other ways. Behavioral observational of baboons have provided enough evidence of their thinking (and scheming) abilities. They may even have different ways of communicating their emotions that are more advanced than the referential language we attribute to them. Such possibility generally exists for all nonhumans with central nervous and vocal systems. And, of course, it would vary greatly among them, depending on the species, and on their place on the evolutionary ladder. To summarize, thinking and language, in whatever elucidations, are intertwined and have evolved together.

The point of departure from what may be called the natural language—the ability to convey thoughts to others, through the usage of various ranges of acoustic waves, cutaneous sense, and visual and possibly other means—to spoken language is around the time of the appearance of humanoids on the life scene. Obviously these creatures had the same communication facilities that their primate cousins had to begin with (and still do—short grunts among different cultures still carry meaning), and these served as a basis for the progression of utterances of thought and the development of language, as physical abilities were being lost and brains were being expanded. The starting point of spoken language, as it is today, whether it was the culmination of neuronal learning or a genetic mutations, is thought to be around the time of the Stone Age, around one hundred thousand (plus or minus fifty thousand) years ago, when art began to appear; though by no means is this a solid proof. The development of any form of language is by no means at odds with the presence of other modalities of communication by other means among thinking creatures, as mentioned earlier.

The Humans

> *Nevertheless, the difference in mind between man and the higher animals, as great as it is, certainly is one of degree and not of kind.*

> Charles Darwin

The most complex event in the history of the evolution of life-forms is the emergence of Homo sapiens—a creature endowed with a bigger brain and with walking and tool-making abilities, though at the cost of the loss of some of the agilities of its predecessors. Obviously this shortcoming would make survival more difficult in the beginning but endowed man with a better grasp of the hardships of wild living and the need for less competitive access to food. They made their foraging ranges more extensive, leading to distant migrations and settlements at various stages where conditions were more favorable. Their survival was also helped by the development of teeth more suitable for meat consumption as well, which saved them from the consumption of unknown and perhaps harmful plants in new territories—animal protein was the safe alternative. The ever-expanding brainpower of Homo sapiens, due to varied and inevitable experiences involving varieties of food sources, landscapes, climates, and natural events, resulted in grouping and cooperation among them and the development of advanced hierarchies in the groups. Branching and the dying off of groups, as part of the natural selection, led to the survival of the fittest, and as such, today's man has emerged at the pinnacle of the evolution of humanoids. Of course these transitions, from the ape predecessors to humanoids and to today's mankind, took about five million years.

In animals, the engine of survival directly drives the (instinctive) demand for food, and at certain times, for procreation. Also, to increase their chances of survival in what must have generally been hostile environments, a preference for group living emerged, the sustenance of which required, to some degree, the development of certain behavioral codes to avoid inflicting mortal damage to one another due to conflicts within the groups. *The different in-group and out-group behavior of chimpanzees, which may involve carnivorous acts in the latter case, speaks to this fact. And perhaps this was where we began.*

For Homo sapiens, the dynamics of survival governed by their emerging brainpower, in time led to the formation of communities, cities, and countries, as well as sophisticated protocols of behavior. But behind it all, the survival instinct, whatever of it had not been tamed by ethics, morality, or intellect, had been chugging along and at work, at times for the most primitive and ***misguided*** preservation purposes. This drive, both at individual and societal levels, was huge and still is, and in many circumstances of human existence, it has served as a demoralizing and at times destructive force within societies, and a devastating one when used outside societies. Wars are of the latter case. Occasionally the power of the primitive instinct of the masses is unleashed under the pretext of protection and survival— exploiting the appeals of (mass) indoctrinated (planted) ideologies or ambitions for international aggressions!

The phenomenon of religion, an amazing, unifying force, appeared on Homo sapiens' mental horizons early on. This creature's innate awe and fear of natural forces, and his miseries and sufferings from their ravages and the harsh living conditions thereof, served as the seed for the creation of religion. And this must have taken shape initially from the soothing utterances addressing the heavens of an accidental charlatan, or an innocent member of the early groups (the would-be medicine man), which evolved in time into the development of great religious philosophies (for their times) and the organized religions of the last couple of millennia. Religion is deep-rooted in the human psyche, since collective consciousness has always been well desirous of it. And the recruited believer masses always followed their religious wont, most times with the aim of getting relief from life's difficulties. However, on many occasions, it was themselves or others of different beliefs who paid a heavy price due to religious dogma, or their leader's hidden material and egotistical agendas.

Societal leaders have throughout history operated similarly by engaging the survival drives and insecurities of the masses, at times for their betterment and on other occasions for personal gains, rendering much destruction and carnage. On the seemingly favorable side of our societal evolution, humans have come a long way in capitalizing upon the use of the survival drive for some measure of a better and longer life (at times this is questionable due to a lack of true quality) for many, albeit at the cost of the dwindling of the very sources of survival that humans and many species totally depend on.

Many times in the past, the world has succumbed to the evils issued, in large part, from the drives of insecure and/or egotistic men in major decision-making positions, and often they were caused by a desire for a rarely justifiable massive gathering of material wealth and power that comes at a heavy cost to others and nature. It is the untamed survival drive of the immature minds, which is behind the fulfillment of mostly mistaken, insatiable "needs." History is telling of such repeating phenomena; it can bring man and animal to the verge of inevitable mass extinction.

Mankind's behavior, despite its potential for equanimity and harmony, makes them as a whole one of the most deleterious creatures that the Earth, in four billion years of natural history, has ever experienced. History is inundated with tales of the horrendous acts human inflict on one another and other beings, undercutting their own long-term survival. Man has been so far his own worst enemy!

Existence is at the mercy of (1) humans' understanding of its fragility, and (2) taming of their survival drives and putting the bridle of knowledge on it. To this end there is a need for quantum jumps in improvement of our collective consciousness in order to bring about drastic changes in the way masses are, or can be, driven. A universal vision for harmonious sustainability needs to take shape in the human psyche, to tame and guide the primitive drives, for the good of all. Will Durant's paraphrasing of Plato's description of Socrates' philosophy of virtue and state pertains to such a wish:

> *If for example, good meant intelligent, and virtue meant wisdom; if men could be thought to see clearly their real interests, to see afar the distant results of their greed, to criticize and coordinate their desires out of a self cancelling chaos into a purposive and creative harmony—this, perhaps would provide for the educated and sophisticated man the morality which in the unlettered relies on reiterated precepts and external control. Perhaps all sin is error, partial vision, and foolishness? The intelligent man may have the same violent and unsocial impulses as the ignorant man, but surely he will control them better, and slip less often into imitation of the beast. And in an intelligently administrated society—one that returned to the individual, in widened powers, more than it took from him restricted liberty—the advantage of every man would lie in social and loyal conduct, and only clear sight would be needed to ensure peace and order and good will.*

To this list we must add the need for *"intelligently administered natural resources"* to save our species—for sure, many others as well—from ourselves, in the industrial age!

The journey to this end is understandably lengthy, arduous, and evolutionary (societal Darwinism), at the end of which, perhaps the realization of Russo and Marx's philosophies in their presumption of man's good nature will materialize. Nonetheless, we have come a long way: our species has formed societies and has—in one form or another—created Leviathans (in accordance with Hobbian philosophy) to control and harness man's individual and collective drives, aligned generally with the ruling body's philosophies and occasionally in accord with societal ideologies. Major parts of the world are suffering from perils caused by devious autocratic, illegitimate, and often ignorant government bodies, which are busy ripping apart and plundering the resources of their countries and populations. There are exceptions, and Western countries are among them: they have made great strides in diverting some of the fruits of progress for the benefits of their masses, at least temporarily, though historically, it has been at great costs to the rest of the humanity and the ecosystem.

Here we need to address the Nietzschean *"will to power"* philosophy and battle with it: Nietzsche has missed the fact that this force has been at work from the dawn of humanity, leading to humans' ways of living, the creation of empires, and in the ebb and flow of human lives and history—their morality, ethics, religion, values, and protocols—all this that he so harshly questions are humans' own making, because *individual* "will to power," in the context of his societal living, has worked out this way, at least in the short period of civilization. And *societal* "will to power" has often also failed to accomplish his imagined goal. Though, giving him some credit, much progress, some civility (which he doesn't seem to like) and morality, and certain international protocols of behavior have likely come out of conflicts and wars.

Societal and individual evolvements, much influenced by mankind's ever-present chaotic cross-encounters within and outside of their living domains, will grope along the Darwinist (societal) evolutionary path to optimize mankind's survival in the context of the eventual ways of nature. The most critical factor on this path is recognizing the limitation of natural resources and their proper sustainable exploitation, rather than mining them, which in final analysis mean population control along with much reduced and sustainable consumption. Therefore, if mass extinction does not happen because of the outrageous mismanagement of the natural sources of survival, or by catastrophic natural events, very likely that means there will be light at the end of the tunnel and the natural evolution of man's society will move toward a global utopian system. The members of the bigger world tribe will have learned to limit their primitive survival zeal to just a desire for a secure and sustainable population and resources, and they will use the power of their gray matter to ensure this fact, which would render the ideal end. And this all has to wait for the emergence of collective brain sophistication, for correct discernment of problems facing life, and their optimal resolutions and solutions. We are pointing to an evolved collective consciousness. And this necessities delving into our understanding of the faculties and the functional aspects of the individual brain; that is, its mechanisms of discernment and perception of the living environment and proper rendering of one's behavior and conduct in life, all which are built on the brain's evolutionary innate and learned constructs (instructions) and its operations. The functional theory of the brain helps to lay out the path: in the next sections, we discuss such a theory, which is based on the discoveries of the past couple of decades in the arena of neurosciences and computational fields.

Intelligent Machinery of Life

● ● ●

Did the sensations of themselves, spontaneously and naturally, fall into a cluster and order, and so became perception? No...putting sense into sensation requires innate knowledge...and because they are a priory, their laws, which are the laws of mathematics, are a priory, absolute and necessary.

PHILOSOPHER KANT, PARAPHRASED BY WILL DURANT

WHILE THERE IS NO UNIQUE definition of intelligence in the literature (e.g., Wechsler 1944; Wheeler 1979; Sternberg and Salter 1982), the essence of it could be summarized as *the application of knowledge to intended goals.* This immediately reflects upon animate beings, as to where in their system knowledge resides and how it is applied. These questions will be addressed in this and the next few sections, as they relate to mammals in general and Homo sapiens in particular. We begin our efforts in the context of an anecdote that lays the basis for what becomes the thesis of the book:

The man enters the casino for the nth time in his life. His heart is pumping faster than usual. He goes straight to the roulette tables, his palm pressing hard on the $100 bill he is holding. His eyes become fixated on the turning roulette wheel. The dealer makes a call for betting, and after a while grabs the roulette ball and twirls it on the side of the roulette. The man's eyeballs start following the ball. After a couple of seconds, the ball drops in a numbered slot. Motionless, he takes a deep breath and looks at the electronic board showing the winning numbers of the preceding runs. His eyes scan the betting board. On this board a crystal weight is sitting on the present winning number. Winners are paid, players start filling up the number board with new bets, and another cycle of the game begins.

It has been almost an hour since he first arrived. You could feel an air of excitement around him from his restless fists and the twitches in his face. It seems like he wants to make his move, but something is holding him back. Suddenly, when the ball is about to drop and betting is nearly stopped, he extends his hand over the shoulder of a sitting player and calls the dealer to put his money on the number eighteen. He walks a couple of steps back while his nervous gaze sweeps the floor. A couple seconds pass, and the dealer calls the winning number. It is eighteen, and he wins.

This is not the first time he has done this. He told me he has to put up a great deal of resistance to his urge of starting to play as soon as he arrives at the casino. He rarely overcomes the urge, and therefore he usually loses. Only a losing trend gives him enough strength to win over the urge and to stand next to the roulette table and focus on the ball and winning numbers for long periods before he makes his winning move. Experiences of this kind are not uncommon to many, though everyone thinks of it as a stroke of luck.

Is what he does really a matter of chance and nothing to do with a person's inner ability to predict the winning number? What if the mind has a great many such capabilities, of which one usually has no awareness, and therefore the telling tales of such an experience are generally ignored! Prediction ability of the brain, of this nature, is only a fragment of the wonders a mind can reveal. Let us find out where they may lie and of what nature they can possibly be and how they possibly can be accessed.

An often neglected clue to such possibilities is from what is known about the blueprint of life, operating in each individual cell in all beings, keeping them alive. This amazing coding in the genes is behind the operations of the cells, which adds up to the development of complex animated beings. It provides for the lifelong sustained operation of the machinery of life, that is, all the required autonomous biological activities of the mill of life, and providing the implements of survival in the face of all oddities! Obviously these codes are the result of the evolutionary process, preserved in the DNA, which are morphed into stupendously complex and well-managed systems for the sustenance of animated beings. The management of these systems would be the charge of the neurons in the central nervous system, driving the operation of the brain to its nature-purposed ends.

Given the unimaginably complex details of how life is developed and how its mostly automated machinery operates to sustain it—all this perfected over time and passed in chromosomes from generation to generation—it would not be unreasonable to infer that undiscovered strands of DNA are likely to hold codes (programs), which could appropriately be morphed into mental abilities that would be beyond our normal consciousness and its demands. However, situations that result from focused and strong demands for favorable outcomes of certain little-understood phenomena—as shown in the story we started with—may entail the requisite mental inputs to cause proper resolution, the process of which goes generally unnoticed and the reasons of which remain obscure. And as such, this, which may in passing loosely be called the sixth sense, will be addressed later in the book.

The Brain, the Mainframe Computer

● ● ●

This new science of mind is based on the principle
that our mind and our brain are inseparable.
The brain is a complex biological organ possessing
immense computational capability: it constructs
our sensory experience, regulates our thoughts and
emotions, and controls our actions. It is responsible
not only for relatively simple motor behaviors
like running and eating, but also for complex
acts that we consider quintessentially human,
like thinking, speaking and creating works of art.
Looked at from this perspective, our mind is a set
of operations carried out by our brain. The same
principle of unity applies to mental disorders.

NOBEL LAUREATE NEUROSCIENTIST ERIC KANDEL

Present-day mainstream computers are essentially
two dimensional. They are based on chips that
must be produced under exacting clean-room
conditions, since any fault can be fatal to their
operation. If they are damaged, they do not
recover. Human brains differ in all those respects.

They are 3D; they are produced in messy, loosely controlled conditions; and they can work around faults or injuries...We may aspire to make body-like machines as well as brain-like computers.

NOBEL LAUREATE FRANK WILCZEK
PHYSICS TODAY, 69, NO. 4 (2016)

PRESENT-DAY UNDERSTANDING OF THE BRAIN attributes a very prominent role to one's unaware (unconscious) mental state in the context of the brain's computational concept. Its operations are considered to be behind almost all requisite life system functions; mostly autonomously, the brain conducts beings' lives, part of which appears to them as the state of consciousness, which is the ultimate indication of the activities of the unconscious, the unsung hero of the brain. To develop a feel for the concept and how it can be identified with, we need to examine some of the mind (personalized brain) phenomena that we are familiar with:

1. Mental events that have always been the subject of attention, curiosity, and fascination are dreams, hallucinations, and so-called revelations, some of which have been the source of agony and unease, such as mental disorders of various degrees.

2. Important and larger-scale prevalent phenomenon, a subject of much bewilderment, has been the common tendency of all cultures—whether clustered, apart, or totally isolated—to develop basically similar mythological symbolisms (Campbell 1984).

3. A less dramatic and more common incident that people of all walks of life experience, which generally goes unheeded, is the sudden emergence of information that they had earlier tried to recall but had forgotten. Among deep thinkers, whether the curious stargazer of antiquity or the researcher of more recent

times, such a moment of revelation, an "Aha!" moment—much elaborated upon by Roger Penrose (1990)—has often been reported. More important cases occur in practices of scientific research when findings occur to the mind either spontaneously or delayed, and occasionally quite unexpectedly; some at times completely outside the realm of the space-time domain of the research environment. The process of thinking does not explain, in a direct way, these late appearances of the answers or the solutions. Discontinuity of the thought process, between systematic thought inquiries and arriving at an answer and/or solution, very clearly points to separate processes, thoroughly outside the domain of (seemingly) conscious thoughts.

Many efforts by many thinkers, from the dreamer medicine men of the early communities and ancient philosophers to today's well-known scientists and philosophers, have been expended to interpret or understand some of these phenomena:

Dreams and hallucinations are justifiably attributed to (the mostly) chaotic synthesis of past experiences stored in the subconscious (possibly more accessible part of the unconscious), and mental disturbances are attributed to psychological and psychiatric phenomena.

Jung (1933) attributes mythological symbols to human's ***collective conscious,*** which according to him appears in certain dreams in order to provide clues for accessing the experiences of the past, which the brain may have been holding, or for forewarning of future cataclysmic events. His Treatise (Jung 1933), along with the findings of Freud (1935) and those from various schools of psychology, is part of the more serious investigations addressing these occurrences as various attributes of the ***unconscious (subconscious)*** and the ***consciousness.***

Discoveries (revelations), always subsequent to an "Aha!" moment, have been attributed by Platonist philosophers to the wellsprings of an imaginary world of knowledge (Gail 1999).

The functional processes of the brain that give rise to hosts of various phenomena, such as discussed, seemingly have not received the scrutiny they deserve. Mainly, behavioral studies (Skinner 1984) and cognition research (Dasen 1994) have addressed the extent of the direct role of unconscious events from the viewpoint of behavioral aspects in the conducts of life. Recent work from the physicist Mlodinow (2011) is a valuable contribution in this regard. A great step toward understanding the mind can be found in the realm of efforts in the field of artificial intelligence—developing machine-driven lifelike activities, which have been of great value in addressing the general functional processes of the brain. With the advent of the foundation of modern computers, around 1940, thoughts of mimicking logical thinking, through the use of such devices, have been on the minds of cognitive and neuroscientists, psychologists and linguists, and experts in other fields as well. Many advances were made early on, though bounded within the limits of the algorithmic treatment of thought processes. However, the presence of domains of thoughts not amenable to logic became the main stumbling block to the further development of artificial intelligence (Turkle 1988). This is partly due to the nature of context-sensitive logical descriptions and fundamentally due to the inconsistency of axiom-based mathematics (Barkley 1939) that would affect computability by mechanical computers, or in general the universal Turing machines (1948). The computational neural net, a construct made on the basis of the layout of neurons and their theorized function in the brain—a connectionist theory—does not suffer from some of the above drawbacks, since there is no need for a precise, logical structure. This approach, embodying an implicit description of a desired behavior through learning, with no need for the declarative logical statements that controlled the earlier artificial intelligent systems, has enabled the emulation of some human-like mental activities. Progress in the field of dynamic robotics attests to this fact.

A reductionist's cursory glance of the biological and neuro-physiological processes that involved the brain would obviously reveal the difficulties, and the shortcomings, of the computational neural net attempts in developing human-like intelligence. However, chances are that with due mathematical implementation of (at least first-order) reductionist concerns, a lot more progress can be made. To this end, great efforts in many fields related to neurosciences continue to help develop a better understanding of the brain's computational functional systems (Edelman 1987).

The above background, ranging from the experiences of the mysteries of the mind to the brain-inspired emulations of machine intelligence, established by the principals of neurosciences and delineated in the massive compendium of the most up-to-date research in the dynamics of neuronal functions (Kandel et al. 2010), has left little doubt that the brain should be considered a neuronal computational machine of immense power. And equipped with an evolution-perfected, innate, and life span–learned knowledge, the inner workings of the brain are gradually being understood.

COMPLEXITY RESOLUTION

Laws of thought are also laws of things,
for things are known to us through
them that must obey these laws.

KANT

Laws of nature and laws of logic are the same.

HEGEL

"...More specifically, the brain is able to learn physics concepts because of its ability to understand the four fundamental concepts of causal motion, periodicity, energy flow and algebraic (sentence-like) representations"

NEW RESEARCH FROM CARNEGIE
MELLON UNIVERSITY (2016)
http://medicalxpress.com/news/2016-04-scientists-brain-
repurposes-scientific-concepts.html

The understanding of many life complexities (governed, in essence, by the laws of physics and nature, some of which are availed to the collective consciousness) requires knowledge of their rigorous symbolic logic formulations and known conditions of the domains of their influences. And this is in order to mathematically express their phenomenal essences. The familiar categories of problems that deploy such formalisms are the class of the initial and boundary value problems, resulting from the application of ***conservation laws of nature (energy, momentum, and mass) to phenomena of interest,*** common in applied sciences and engineering. With few exceptions, all such problems prove only amenable to numerical solutions requiring complex manipulations to convert the formulations to a form that can render them solvable by digital computers. These final forms are generally sets of parametric linear-appearing simultaneous equations that hold true for small values, or variations, of the state variables, in a discretized space and time domain. Known initial and boundary conditions, and the assumed or known behavior of the phenomena (constitutive laws) in the solution domain, make the problems deterministic by rendering all parameters known. Inherently, the number of equations and unknowns would be equal, ensuring unique and nonredundant results. Progression of the solution in increments of space, time, and related boundary conditions, either iteratively or directly, provides the proper range of expressions of phenomena under consideration. Digital computers, requiring such logical and explicit mathematical formalisms, have made the solution of very many complex problems, such as the building of civic structures to the launching of space vehicles, possible.

As discussed earlier, a fundamentally different approach for solutions of certain classes of complex problems is the computational neural network method (McCullock 1993), which is ***inspired by the conceived information processing of the brain***. The neural net solution approach is well suited for a host of problems where, mainly, the complexity (posing varying degrees of difficulty for alternative approaches) defies the development of mathematical formalism. Also, neural nets avoid the fundamental incomputability flaw inherent in the axiomatic mathematics on which digital computations are based. In the neural net method, a presumed embedded (implicit) mathematical formalism, and the initial (condition) assumptions (the neural node connection weights), and the boundary (conditions) values, which are the complexity-sensing nodes inputs, make the problem mathematically well posed. And the solution approach starts with the training of the neural net, which can be likened to a kind of verification/validation of its presumed computational formalism. The process requires iterative signal-weighting manipulations and calculations. The calculation formalism—at least, in its simplest mode—is akin to the setup of a number of parametric simultaneous equations (Schad 2016) with varied degrees of coupling among variables, which are solved by the physical construct of the net or, alternatively and prevalently, by the digital iterative solution of a net-equivalent formulation. ***In the latter case, the mathematical formalism (of such nature) is very likely the underlying problem-resolution layout of*** *the physical neural network.* The parameters of these equations, referred to as "knowledge" values, are the values of the connections' weights. Any changes/modifications in the knowledge values—stored in the network as results of prior iterations, or trainings toward completion of the training—require the deployment of a (mathematical) learning rule, which can perhaps be very loosely likened to a certain constitutive law (though somewhat arbitrary for nets) in the traditional problem-solution domains. *Again, the solution process, essentially the training of the neural net, enables such devices to provide solutions to similar classes of problems that are either very difficult or impossible to solve otherwise.* In comparison, while the traditional mathematical (numerical) approaches rely on the availability of rigorous logical formalisms to render problems solvable, in their discretized domains, through algorithmic resolutions, the power of the neural networks lies in their semblance of operating in the inherently discretized (*physical or mathematical*) domain of neural nodes, resolving problems, though without any use of a priori mathematical formulations.

Problem solving, in the realm of man's conscious mind (personalized brain), begins with the onset of thinking, for conceptualization and/or formulation (mental scripts). For many cases of mentally challenging problems, the solutions are generally preceded by an "Aha!" moment foretelling their manifestations. Undoubtedly this is indicative of a discontinuity in thinking, which very clearly points to separate backend processes in the brain, thoroughly out of the domain of (seemingly) conscious thoughts. Obviously, many a problem, including the ones benefiting from common-sense solutions, with no indication of the *thought-to-solution* process, relies on the brain's instantaneous accommodation of solution demands. Such dynamics can reliably be attributed to the brain's computational operations, especially when addressing beings' awareness of the world, a phenomenon that is only conceivable in terms of mental models in the brain, considering the electrochemical nature of the sensory signals that it receives.

The brain computational approach, in all likelihood, must be from a ground-level computational perspective, similar to the (brain-inspired) scientific neural network (Schad 2016); their network structures, and the latter's dexterity in solving complex problems, bears this estimation. Obviously many facets of the brain computer—its operational details, along with very legitimate fundamental questions from reductionism points of view—will remain to be worked on, perhaps not ever to be completely answered. Nonetheless, the brain is fundamentally conceptualized here in the likeness of the scientific computational neural network (recent experiments discussed later add much credence to this presumption). Incomparably more powerful on this account is its massive parallel processing (perhaps even quantum) computation ability. The brain neuronal network is a survival-perfected/tuned physio-biological construct, which has evolved—genetically sculpted and configured in the face of nature's complexities, which beings have experienced in the evolutionary time span, and unequivocally scripted for the survival of species. Kant was on the right track to suggest the presence of a structure inherent to it (not a tabula rasa). At this juncture it should be added that contrary to belief, among some philosophers (e.g., D. C. Dennet), that genes have no more role past the development of the brain neuronal structure, gene expressions contuse in manufacture of neurotransmitters at synapses—rendition of Post synaptic potentiation and hence firing that problem resolution depends on it.

Having inferentially established the nature of the brain computational machine, we can assume that the complexities (problems) related to beings' lives, internal or external, would be implicitly algorithmized in the domain of the synapses (the resolution of complexities) as multitudes of parametric simultaneous equations. Rendering what may in essence be called **the *equations of life and living*.** And for many such equations, the parameters (the nodal weights, the knowledge) have been naturally worked out, and their autonomous solutions bear most of the requisites of life and living for each being. A viable argument for this claim can be made by considering the deformation of a piece of metal subject to heat: the metal does its thing by obeying the laws of nature. Governing laws of this process are known, and the results can be observed. While to understand and predict the observed phenomenon of the heated metal expansion, we need to solve the diffusion equation (heat law) using the constitutive behavior of metals in the mass of the metal, given the properties of the metal, its initial and air (boundary) temperatures, with due considerations for the convections and radiation that are taking place at the boundaries, all for finding out how it would behave. The brain—being a computational device orchestrating complex biological systems conducts, unlike a piece of metal, not the system itself—has to have the formulation of every phenomenon it handles hacked into its construct, which are the algorithmized versions of the governing laws. And this would be true for all laws of nature affecting life: ***the brain is ready to render solutions, immediately or by trial and error, when prompted by complexity exposure stimuli.*** The latter case arises in the conscious management of life, which may give rise to the initially undetermined mode, which is the subject of next section.

COMPLEXITY SOLUTIONS

> ***The laws of Nature are written in the language of mathematics.***
>
> GALILEO GALILEI

Brain computational outputs are the results of the execution of instruction upon life environment (internal and external) demands. They drive the body's bio-physiological responses. Obviously, much of these processes remain obscure to the consciousness, but we become at least partially conscious to some of it. The latter happens during exposure to problems; in thinking, or seeking solutions to complex problems, when the "Aha!" moments occur. Thought-engaged inquiries are all in the context of the conceptual general initial- and boundary-value problems, where—partly aware—the trial-and-error learning process toward final resolution takes place. What is learned of the external world (experiences) at every step in time, while mostly enhancing the computational powers of the brain—forming neural patterns and structure development—prepares the "initial conditions" for the next step of the solution in response to varying "boundary conditions" and the procession of time. It's important to note that despite the seemingly intentional (conscious) engagement of the individual in the outcome, all operations are autonomous in nature deep down. From a grander perspective, ***the principle of causality*** has for long proved this in the eyes of many who have pondered the question of free will.

The role of the antecedent events in shaping the succeeding ones can best be portrayed in human life. Considering the development of the very initial conditions at the start of life (obviously differing by DNA gene contents and their rendition of different physiognomies and brain constructs) for each individual, followed up by its changes as a function of varying boundary conditions and progression of time (different exposure environments), vast variability in outcome are warranted, as evinced in differences among humans.

From the operational end, the trial-and-error approach involves the deployment of the being's biophysical feedback mechanisms (the senses and related physical implements), which supports the calculational operations of the brain, a process that determines neuronal signal weight distributions—*the web patterns* (constructs). And the solution event is equivalent to solving sets of simultaneous equations of a certain number of unknowns—*equations being the neuronal patterns, the implicit algorithmic expression of the brain-resolved complexities under consideration.* In such likeness, depending on the degree of complexity, the number of equations and unknowns vary, which implies engagement of varying segments of the neuronal domain. The solution proceeds as parameters are appropriated by trial and error, and of course any available knowledge (predetermined neuronal weights) is also used. The output of the brain system, depending on the degree to which solution parameters are correctly determined, can be imperfect, perfect, or redundant, which determines the state of consciousness and the ensuing (input/output) interactions with the surroundings. As mentioned in the preceding section, immediate solutions benefit from the availability of the infinitude of a priori learned *patterns*, some issued from genetic heredity and some configured earlier during lifetimes. These facilitate solutions of continually posed problems, engendered by ever-changing life (*boundary) conditions*.

Although rigorous direct testing of the laid out (brain) mathematical approach to complexity solutions has to await further developments in the understanding of brain's functional operations, one may be able to find anecdotal hints that may point to the robustness of our thesis. One such case describes the results of an LSD experiment, which will be discussed in another section. The gist of it is that the loss of ego, reported by subjects in the experiment, which coincides with fMRI evidence of massive engagement of the brain's network, is also reported in practices of high meditative states achieved by yogis. Clear inference from how, in the latter case, such a state is achieved—by placing a huge futile resolution demand on the brain—strongly supports the computational resolution/solution concept followed in this book.

In summary, life's complexities are sensed and discerned (resolved) in the dynamic domain of the nodal spread of the brain's neuronal network; that is, they are implicitly algorithmized for solutions (properly assimilated), forming, in essence, beings' governing equations of life and living, which are then solved— practically in the way of many known scientific complexities, which are algorithmized in the context of initial and boundary value problems in their discretized domains as sets of parametric simul- taneous equations, that renders them solvable by traditional com- putational operations, directly or iteratively. The results in case of the brain, render the perception of events and realities of life. The computational operations of the brain define two mental states: the unconsciousness and the consciousness—the unaware and aware states—and the latter describes the aware, interactive living pro- cesses involved in charting life's path, following nature's causality principle.

Perception of Reality

● ● ●

Sensation> Resolution> Solution> Perception

*The mind is not an agency that deals
with ideas, but it is the ideas themselves
in their process and concatenation.*

BARUCH SPINOZA

The world is my idea.

SCHOPENHAUER

GIVEN THE REALITY OF THE computational machinery of the nervous system, it has been inferred that (1) the brain must be behind the resolution and solution of an incredible number of life's complexities, at the service of its sustenance (physical, biophysical, behavioral, etc.), and (2) it also simultaneously has the mysterious and fundamental task of perceiving the world in which beings need to be (at least partly) conscious of. And since beings are conscious of themselves and their environment, and since all life events occur in such domains, their perceptions must be in the realm of ***representations*** of all realities. This is because perceptions are based on afferent sensory pulses (electrochemical spikes of various intensities at the synapse level), and the ***representation can only be computational created***—the density and volume of data preempts much role-play of the memory system. As such, life is perceived in this simulation (representation) of the real world that the brain creates, and its events are reconstructed by the innate, embedded evolutionary as well as learned constructs (neuronal patterns rendering specific computational instructions) from the integrated sensory stimulation inputs.

It is important to note that the simulation argument here has little to do with the simulation argument raised by Bostrom (2003). The latter includes a hypothesis for the possibility of our world being a sophisticated computer simulation generated by a more sophisticated technology than our own. The simulated world I refer to is a close brain representation of the external realities in which humans are sustained.

Considering the fact that the perceptions of our environment happen because of the brain's processing (simulation) of the pulses it receives, the mechanism and where the experience of perception occurs remains a mystery. However, perception in general is taken for granted (except perhaps for the visual experience). Maybe the reason is the fact that other senses, which use material contact (matter molecules), provide a sense of continuity with the perception of the body itself. However, the question of the nature of vision, and its display medium, has always been a source of amazement and wonder to curious minds. While the neurobiological aspects of vision are well understood (Schwartz 2009), the puzzle of vision perception has defied any attempt of resolution so far, as the author understands. In the following section, we attempt to divulge the nature of vision and how and where it is perceived. And since the experience of vision occasionally invokes some thought—both being outputs of compu-tations—and since their mechanisms have common roots, it will be addressed as well.

Vision and Thought

> *And the tongue speaks in accordance with*
> *what is seen or heard. But when the brain*
> *is still, a man can think properly.*

Attributed to Hippocrates, fifth century BCE

> *Every idea, however abstract, moves the*
> *body in some degree, however unseen.*

Philosopher Thomas Hobbes

As argued earlier, the brain is the site of the mental enterprises, a consensus reached since the age of enlightenment, which is now believed to encompass extensive computational facilities and fac-ulties that process senses-relayed information, as well as those of any possible mental codes, like the "I" (internal) languages that Biolinguists have hypothesized (Chomsky 2007). Experimental evidence in neurosciences, based on direct or indirect observations (e.g., fMRI and encephalography), point to somewhat of a modular segmentation of the brain in relation to the senses and other mental capacities. Considering the vastness of the computational neuronal network of the brain—one hundred billion neurons, each connected at ten thousand synapses—an apparent segmentation (appearance of modality) should be expected if engagement of all or a large part of the neuronal computing network would not be necessary;

evolutionary checks-and-balances processes must have resulted in a computational economy that would establish the division of labor for the brain. A basic similarity between the computational principles of the brain-inspired scientific neural network and the brain neuronal net allows an important deduction: in the former, the more complex the problem, the more extensive number of node layers and nodes are needed; in the brain neuronal net, hierarchical grouping and plasticity facilitates this scheme. The *similitude principle* inspired by Lord Rayleigh (1915) adds credence to this assumption:

> *I have often been impressed by the scanty attention paid even by original workers in physics to the great principle of similitude. It happens not infrequently that results in the form of "laws" are put forward as novelties on the basis of elaborate experiments, which might have been predicted a priori after a few minutes' consideration.*

As is observed, the size of brain modalities is determined by the multiplicity of data. This is evinced by fMRI evidence that shows engagement in different locations and different sizes of brain segments involved for various senses. For example, in the case of vision, massive data pulses are sent from the millions of photoreceptors in the eyes, which expectedly would engage a larger capacity of the computational facility of the brain.

Optimum perception for beings requires the functioning of all senses—four out of the five senses require contact with matter (solid, liquid, or gas) in order to be activated. The vision sense, unlike other senses, is a remote sensor: it is the wandering of light, returned from the embrace of the physical world that catches the attention of this sense. The absence, or a deficiency, of any of the senses affects perception to varying degrees: a very dramatic one would be due to a complete lack of vision. Evidences from the congenially blind suggest heavy dependence on, and use of, other senses, especially somatosensory, for some measure of compensation for the defect. Anecdotally, a dramatic scene of such a case was seen by many in a broadcast of the meeting between Helen Keller and President Eisenhower, where Helen was permitted to touch president's face to "see" him, as she put it (video available via YouTube). Considering the scarcity and lack of simultaneity of touch data, the congenially blind require training and practice in early life for managing their limited perceptions. Though blind persons describe "seeing" the object they touch, ***they do not experience vision-like perceptions, neither in waking hours nor in their dreams.*** Obviously this shortcoming in the lives of the fellow human beings has always been on the social conscience, and efforts have been made to curtail it; new possibilities have already begun and are continuing (Bach-y-Rita 2006).

Visual simulations—what we sense as seeing—are based on information received from millions of retinal photoreceptors sensing light from the environment, mostly supplemented by information from other senses (like touch) and use of the embedded experiences of distance (relative locations). The computations required for simulation engage a major part of the brain to create images of the external reality. The fact that simulation takes place is also suggested by the ability to imagine faces and settings during waking or sleep states. The projection is based on (1) the availability of the massive computational (parallel processing) power of the brain's neuronal network and its imbedded resolution and solution constructs (patterns) (Broad 1978; Juyang 2012), and (2) the immediate accessibility to memory and sensory data—enormous simulations for scientific and engineering tasks are routinely performed by comparably much less powerful computers. Obviously, the deliberation can only be firmly established if the underlying (mathematical) formalisms for such simulation abilities in the brain can be understood. An important step toward this goal is to assume that, by virtue of the ***similitude principle,*** the (imbedded implicit) underlying mathematical formalisms of the operations of the scientific neural nets (fundamentally) also holds for the dynamical computational operations of the brain (Arbib 1989). This puts much of the thesis of the book in proper perspective!

In the realm of our perception of the world, whatever way it is achieved, the questions of its ***accuracy*** and ***reality*** are known points of contention among philosophers. It is true that perceptions are the experience of (subjectively) created representations of the world, and the objective reality of *"things in themselves"*—as philosophers like to put it—cannot be known. But perhaps solace can be sought in the continuous, instantaneous, and mostly autonomous verification/validation process (sensory-motor integration) that takes place within the accessible, immediate environment in which beings exist, because it ensures the "reality" of things in terms of their own reality—***the world is as real as we are***. And, adding to that, is the overall semblances of experiences of subjective perceptions among interdependent species, which adds further credence to their perception of a world reality the same as the reality of *"them in themselves"*: the ***world is as real as existence***. Therefore, beings experience the reality of things to the degree of the limitations of their senses, and, as such, the long-lasting philosophical questions of the ***realty of perceptions*** are settled on the side of the "Naïve Realist" philosophers (Searle 1999).

The understanding of any computer simulation requires displaying results on an interface, a venue. Considering the "mind" (an abstract notion that past philosophers and present-day scientists still struggle to define) as the interface, relegates understanding to a philosophical dead end. Suggestion of the thought interface, though still vague, is more rational since it is through thought that awareness comes about (Chalmers 2009). Chomsky (2007), in the explanation of his biolinguistic theory, refers to two separate interfaces: one, the *"thought system,"* where the language mechanism synthesis of structured expressions finds interpretation in it, and two, the *"motor sensory system,"* which renders vocalizations. Vocalization, whether it is referential as in the calls of animals or verbalized as in humans, is a carrier of audible messages of evolutionary value.

The vocal instrument, which will be referred to as the biomechanical *utterance interface* of the motor sensory system, could also be a natural outlet for other motor (efferent) signals. As such, it would serve to display the complete and dynamic brain processing results of the senses intake of posed living environment situations. During waking hours, the senses are engaged in nonstop collection of environmental information (stimulations), resulting in concurrent brain simulations and its consequent stimulation of the sensory motor neurons (the efferent signal outputs). And therefore, the only way to account for the limited vocalization that is the prevailing quiescence among beings with vocal instrumentation would have to be inaudible utterances. This means the utterance interface acts as a major interface between the brain and the external world. Many of the comprehensive simulated constructs and computations of the brain are interpreted into audible and inaudible utterances due to the activations of the motor neurons— some also show as facial expressions. Exceptions to generality are cases where either vocal instrument is lacking or the motor signals are very weak. In both cases brain outputs are diverted to murmurs in the muscular and skeletal system. For speech mode, the motor signals can be diverted to a specific member such as hands (talking with hands which we all do, to different degrees), eyelids, or eyes. The latter is the case for the astrophysicist and black hole theorist Stephen Hawking; his thoughts are sent to the eyes, which act as an interface. Using EEG and speech synthesizers, he can convey his thoughts to audiences.

Inaudible utterances also take a major share of brain outputs, though this is violated somewhat by loud thinkers, who sound off their thoughts. Species survival must have played a big role in the partitioning of audible and inaudible utterances, as the dynamics of social life became an evolutionary pressure. The intensity of motor neurons excitations, related to making the gear shift between audible and inaudible vocalization, must have had to adjust to satisfy these utterance demands. At an observable level, differences between the verbalization of thoughts among different cultural groups are noticeable where one observes more measured, and less frequent, vocalizations than others. The preceding reasoning lays the ground for the following:

Utterance (vocal) interface is the unique venue for outputting the results of the brain's simulations of the real world, as well as other related computations, in audible (referential or verbal) language and inaudible language (thought).

Quoting the statement "Have you ever stopped talking to yourself?" from one of Professor Chomsky's recorded talks provides anecdotal support for the above claim, first by hinting at the reality of inaudible utterances and second by stating that thoughts can always be verbalized.

Aside from language—genetically coded or neuronal learned—a major computational task of the brain, as referred to earlier, is creating visual simulations. The current understanding of the nervous system and of neurocomputational concepts leads to the assumption that the environmental information received from the eyes is processed and somehow rendered into a visual perception. This assumption needs to account for a perception venue or platform—a LCD or a LED in the lexicon of electronic visualizations, as we have come to know—to receive the computational results of the simulations. The utterance interface serves this function for thought and speech computations. However, despite the advanced understanding of the biophysiological and optical aspects of vision, the mystery of "how and where one sees what one sees" remained an enigma until a recent discovery (Schad 2016).

The idea of pattern processing of the skin—perhaps motivated by the nature of perception in blind people—became a subject of study at Princeton University in the "Cutaneous Communication Laboratory," founded by Professor Frank A. Gildard and Dr. Carl Sherrick in 1962. The initial work at this laboratory piqued the attention of Dr. Bach-y-Rita (1968) and colleagues, who laid the foundation of what is now called "sensory substitution systems," based on observations and his understanding of the plasticity of the brain. His review paper (Bach-y-Rita 2006) provides a complete perspective on the state of the art of such research at the time and specifically of his work on tactile vision substitution systems (TVSS). The essence of such systems is the reactivation of the idle, perhaps partly atrophied, parts of the neuronal net of the brain, or the engagement of other available parts, to perform the task of perception when massive sensory data becomes available. In TVSS, a vibro-tactile, or an electrotactile, interface, connected to an environment-detection system consisting of a camera and necessary electromechanical gadgetry, is placed on the skin or on the tongue to transfer appropriate pulses to the somatosensory system.

As shown in tactile vision substitution systems and related works (Bach-y-Rita et al. 1968; Keysers et al. 2004; Keysers and Gazzola 2009) and other experiments (Juyang 2012; Janglova 2004), cutaneous sensations could invoke vision-like perception. This happens if a good measure of data simultaneity and numbers are provided. The fact that congenitally blind people normally have no visual sensation at all (Keller 2011) is not contradictory to the claim. Their vision-like sensation develops only as a result of the generation of abundant simultaneous pulses from an interface patch on the skin or on a smaller area of the more densely innervated tongue. Obviously no "normal" contact of the skin generates the signal density and simultaneity that such interfaces provide. The intense TVSS computational demand on the brain likens itself to that of vision (Ortiz 2011). However, despite the vision cortex being fooled into heavily engaging with the demands of processing the intense TVSS signals, the subjects should feel as though their hundreds of affected nerve endings are touching something (matter) to resolve its configuration, in the same way that they would when touching the surface of an object, since the afferent signals are of tactile nature. Therefore, the expectation would be that they should experience the perception of the patch and its details. ***However, the fact that the early blinded, and ones with blocked normal vision (in various tactile experiments), claim experience of vision-like perceptions speaks well to the tactile nature of vision sensations.*** The brain plasticity role-play, other than the substitution of the function of any present atrophied parts or the management of the extent of brain modality involved for processing, has no bearing on the nature of these perceptions.

Brain circuitry engagements in the perceptions of the simulated world are evidenced by fMRI imaging during various experiments and well indicate the (computational) backend similarity of vision experiences, whichever way stimulated, through photoreceptions or cutaneous (tactile) sensing. However, the renditions of perception experiences require engaging a biophysical interface for the brain computation process results, that is, a display venue for the "motor (efferent) signal outputs." This fact proves the commonality of the interface among all (normally sighted and blind persons). The likeliest candidate for vision experience (by eyes or TVSS), though seemingly strange, is the same utterance system. This possibility is also borne by the fact that vision experience is a recitation of tactile sensation, which mostly occurs autonomously and inaudibly, like thought. Therefore, sceneries are perceived as they are uttered, inaudibly or audibly, consciously or otherwise. This is by no means any different "in nature" from the perceptions of a congenitally blind person, with the exception of the incompleteness of the perception due to scarcity of tactile signals in comparison with visual ones. Had the congenitally blind person had millions of nerve endings to touch all the nooks and crannies of his environment, he would have the same perception as anyone with sight. This is why TVSS creates limited visual-like perceptions, as opposed to those created by eyes, which leverage the sensory input of light to precisely relay massive simultaneous data, something that normal touch is not built to convey. Hence, TVSS subjects experience a hazy black-and-white simulate reality. It is important to note that even healthy eyes miss much of the vast color spectrum of light, and as MIT Professor Frank Wilczek put it, *"...what we perceive as color is a crude hash encoding that lumps the power spectrum into three bins..."* Accordingly, touch signals also lump all frequencies together, and the hash encoding delivers only black and white.

Given the above reasoning, the following can be summarized: *What is perceived as the vision experience is in reality the inaudible, and occasionally audible, recital of environmental simulations based on the integration of a massive cutaneous-like sensing of the environment, and that vision perception, in nature, is like cutaneous perception, and it is the same, in essence, for all, blind and otherwise, which in the former is understandably very limited.*

The very comprehensive and detailed experimental work of Keysers et al. (2004) and others (Keysers and Gazzola 2009), under general investigation of mirror neuron activities, examined "tactile sympathy." It describes the activation of the same areas of motor neurons in a group of subjects watching a video of a second group being very lightly touched on the skin. Combined fMRI and TMS evidence of such sympathy is shown in the work of Alaert et al. (2009). I believe these results provide very strong experimental support in validation of the tactile nature of vision, as presented in this theory. We also can cite a great experiment of nature for support of our thesis: bats find their prey using the echoes of their generated sonic waves and may be said to be seeing what they aim to eat—they are touching it remotely through the modulated returning waves (vibration of air matter molecules). It isn't direct contact but remote touch sensing of matter. In case of vision, the modulated returning electromagnetic waves (light) from objects, perform similar function, but relaying intensely more knowledge about the world out there.

It is noteworthy that the tactile theory of vision also explains the phenomena of "blind sight." This phenomenon relates to the ability of people who are visually blind but can distinguish between objects and shapes, and they are not aware of their limited vision. In the context of our theory, the vision signal, despite its intensity, can only receive limited interpretation in the brain, due a loss of network capacity—the vision cortex is heavily damaged, while the eyes are functioning properly. And because the objects are remotely sensed, unlike direct touch, some training is required before perception can be experienced.

Given the soundness in the theories of thought and vision, additional validation experiments would likely further substantiate them. Experiments monitoring specific vocal motor neuron activities in different groups of subjects, some with normal sight and others with congenital blindness, during speech, thinking, writing, and visual engagements (vibro-tactile touch in the case of blindness), as well as recording of vocal vibrations during conscious thinking, are a few that could easily be performed.

VOCALIZATION AND LANGUAGE

The cutaneous sensing of natural sound vibrations (on skin) and its feedback define sensory motor integration in the process of computational simulation renditions in the brain. This operation may have originally espoused the development of the biomechanical vocalization (utterance) apparatus. Of course, certain mutations favoring the survival of a large branch of species must have allowed it. The audible utterances would reflect some aspects of beings' perceptions of their surroundings and their experiences, in time developing into what is referred to by linguists as referential language.

It has been generally accepted that nonhuman species with advanced nervous systems probably possess some capacity for thought. Of course, within the context of computational brains, this ability would be expected, which implies its perception through inaudible utterances. The development of thought among beings must have preceded vocalization ability, since it is the efferent (computational) result of interactions with the realities of the existence of a higher animal.

Since animal species have limited referential language with which to convey their thoughts, we must surmise that perhaps communication among their social groups occurs by means other than their limited utterance capabilities. Obviously such abilities, if present, would vary greatly among different species, depending on their evolutionary paths. A mutation in the vocal apparatus of a certain branch of humanoids some fifty to one hundred thousand years ago proved highly conducive to the origination of verbal language. However, it is very likely that communication with sign language precedes development of spoken language in early Homo sapiens. The fact that almost all humans use hands while speaking, point to this fact, that is simply caused by the issuance of motor signals throughout the body.

The States of the Brain

● ● ●

Unconsciousness is the original and natural condition of all things, and therefore also the basis from which, in particular species of beings, consciousness results as their highest efflorescence, wherefore even then unconsciousness always continues to predominate. Accordingly, most existences are without consciousness, but yet they act according to the laws of nature.

SCHOPENHAUER

UNCONSCIOUSNESS

SENTIENT BEINGS, IN GENERAL, ARE barely aware of the operations of the autonomous machinery of their brains, which can be imagined as a kind of "cloud computing." Of course, on the face of it, all humans attribute intelligence to the brain, and many among them identify it as the system responsible for the maintenance of the basic necessities of survival and whatever else they find themselves in pursuit of (albeit very vaguely though!). However, among the deep philosophical thinkers and scientific minds in different fields, it has been established that the forces of nature—playing a game of chance, over millions of years, perhaps—from the dawn of the development of the nervous system, have been sculpting the structure and the machinery of the brain (its computational infra-structure) to sustain the inner workings of life and management, some reflected into consciousness, to insure the species' survival in the physical world. A major inference in the process of the develop-ment of the brain's machinery is that all the laws of nature must have been burned into the brain's neuronal web in order to enable the computational modeling of the living environment, based on the sensory pulses it receives, as well as the computational renderings of all other perception experiences.

In spite of our physiological knowledge of the innards of the brain and its likely method of operation, one cannot be aware of the inner working of the brain machinery, that is, its programs and their executions and complexity resolution/solutions schemes, as well as some of the stored operation results, which relate to living expe-riences. *All which is involved in the brain's (backend) processing operations, and not availed to consciousness, is referred to as our unconsciousness,* as depicted in the flow diagram of figure 1.

From the evolutionary standpoint, the template for the compu-tational construct of the brain is issued from the gene transcriptions expressions, the same way the physiological structure is config-ured. However, lifetime experiences add to the brain constructs by modifying it by virtue of its plasticity, which interprets itself in the expansion of unconsciousness and consciousness, which also implies further learning abilities. Darwin's theory of evolution through natural selection is all encompassing in that each biological form needs to be armed with its own survival means, which has to do with emergent unconsciousness/consciousness in various degrees to begin with—at least in higher mammals. Considering the biological adaptability of many beings, such as changing color, cover, etc., the absence of such possibilities among human species must have been compensated for by the expandability of their unconscious powers. To this point, Darwin believes *"It is therefore highly probable that with mankind the intellectual faculties have been mainly and gradually perfected through natural selection."* Therefore, in the case of humans, the development of the unconscious has benefited from biological survival-based adaptations in its animal predecessors and, from the Stone Age onward, from gradual intellectually based growth, both becoming a genetic predisposition. What this implies is that learning during the lifetimes of Homo sapiens has helped further evolution, probably most effectively in the case of humans.

Considering what has gone into the structure and construct of the brain, over its evolutionary time spans, it can be inferred that the unconscious may have some imprints of consciousness from the evolutionary path (other than what is obvious) from our predecessors; that is, the brain—via DNA gene expressions—may carry time-marked information of collectively experienced events or conditions, in the processes of the forming and reforming of the biological species in their evolutionary march. Given such unorthodox takes of the brain and its functions, it is possible to conclude that the unconsciousness, which is placental to every form of higher life, is the depository of many of the secrets of its evolutionary history. As such, these secrets are not only limited to information needed for the formation of just the next mostly similar life-form; rather, they also contain clues to all natural laws as well as some of the other evolutionary path experiences and data.

Such statements of speculative philosophical nature—perhaps partly at odds with scientific discounting of experience inheritance—are partly rooted in inferences from the accepted premises of the inheritability of traits, personality, and psychological disorders by the scientific community.

Consciousness

The phenomenon of consciousness has been the subject of philosophical thoughts, debates, and intrigues during historical times. Important schools of thought are now convergent on the premise of the consciousness of higher-ranking animate beings, and it is defined as the state of awareness of one's whereabouts in life and of one's awareness of interaction in environmental and societal settings. However, the fact that consciousness is subjective (in the sense of differing beings answering the question "being me" or "being you" and all the individual takes on the world) poses problems.

From this perspective, philosophers now divide consciousness consisting of *easy* and *hard* problems. Paraphrasing Chalmers (2014), easy problems are defined as explanations of the objective functions associated with beings' behavior in general, explainable by mechanisms in the brain. Hard problems are associated with the subjectivity of the mind and the world—how can physical processing render it? Various schools of thought generally are categorized as materialism, which denies the hard problem; dualism, which places consciousness outside of the physical realm (somewhat like idealists); and panpsychism, which professes consciousness to be the nature of matter at particle levels, which somehow gives rise to consciousness in beings. And, of course, there are other lines of thinking that consider consciousness a result of the collapse of quantum or gravitational waves. There are no robust philosophical answers to the hard problem of consciousness, though in sciences, empirical theories are being formulated in the search for it. And the approach presented here, being also scientific, may resolve the question altogether:

From the perspective of the computational brain, *consciousness is the partial manifestation of unconscious brain processing (execution) of the continuous and impromptu mental scripts (programs), dictated by the engaging complexities and the necessities of life and living. The operations call upon all of the lifetime's learned and genetically inherited imbedded knowhow (instructions), which are expressed as distributed topological patterns and physiochemical insignias in the constructs of the neuronal web of the brain; subjectivity is at the heart of the process!*

What is known as thought or thinking is in essence downloads of the steps of (antecedent) brain processing outputs to the utterance interface, an emergent phenomenon. The internal dialogue, which one occasionally engages with, is due to insolvency of complexity scripts (programs) in the brain, which engenders selection and decision operations in conjunction with the data and rules (a priori checks and balances) from within and without.

An amazing aspect of brain processes has been the rendering of mankind's apparent ability to introspect, at the helm of which is thinking about thinking. Introspection, along with the other baffling and amazing elements of consciousness such as judgment, intuition, and inspiration, to name a few, are manifestations of checks and balances operations in the brain, or validation, engaging various in-place value system scripts. The results define the disposition of humans internally and externally. This signifies the critical role of societal and personal moral and ethical codes, as well as mental abilities—mostly nurtured in early life—in charting the course of human behavior.

Generally, the interactive process of the consciousness, through ever-present trial-and-error learning (interactive engagement of senses), lends itself to the enhancement of the neuronal patterns and hence further development of the computational abilities of the brain—that is, an increased aptitude in problem resolution and solution. Such brain processes have been the main engine of the development of humans' rich cultures, mythologies, religions, philosophy, and sciences. Unfortunately, this development in thinking ability and all that is learned in a lifespan leaves no genetic imprint of use to the immediate progeny (a priori inherited knowledge scripts are developed in evolutionary time spans). However, man has found ways to preserve generational knowledge gains for the benefit of future generations. ***By laying down pathways of mental engagements in culture, philosophy, and sciences, the domains of knowledge and learning gets transferred and expands from generation to generation***

In the light of the preceding takes of the states of the brains of (biological) animate beings, the question arises whether non-biological animate beings are also conscious! The answer is a qualified yes, and the difference in the nature of consciousness between the two cases is not, in all likelihood, in kind but in degrees: The logic is that such artifices also manifest the outputs of their "computational brains'" execution of embedded instructions— leaned or otherwise. Heir's is obviously limited consciousness and its degree of sophistication would depend on the number of processors that can be deployed in the network of their "neural brains", and the learning algorithms. With the upcoming era of atomic transistors (reaching the limit of Moore's Law), Robots of high degrees of intelligence and learning ability, emulating much human like behavior, can easily be conceived of. Obviously the gap between the Robotic and biological consciousnesses will always remain immense, but we should not be surprised to observe similar displays of many facets of human consciousness in future humanoids.

Anecdotally, considering what earthlings are presently achieving in space exploration, it is not hard to surmise that in this vast universe of ours more advanced civilizations of earth like planets—a statistical possibility-- could have already developed very advanced Robotics, among other imaginable achievements, long time ago (Fiction writers must be way ahead on this). And their

Enterprise Space Ships (equipped with atomic reactor power generators and plasma propulsion system), piloted and manned by humanoid, would be cruising different constellations and gathering data by sending self annihilating—due to the forbidding needed propulsion in gravitational escapes—humanoid missions to various planets (Roswell story, if there be any grain of truth in it, could at best be just a rare failed scenario in such humanoid visit of our planet).

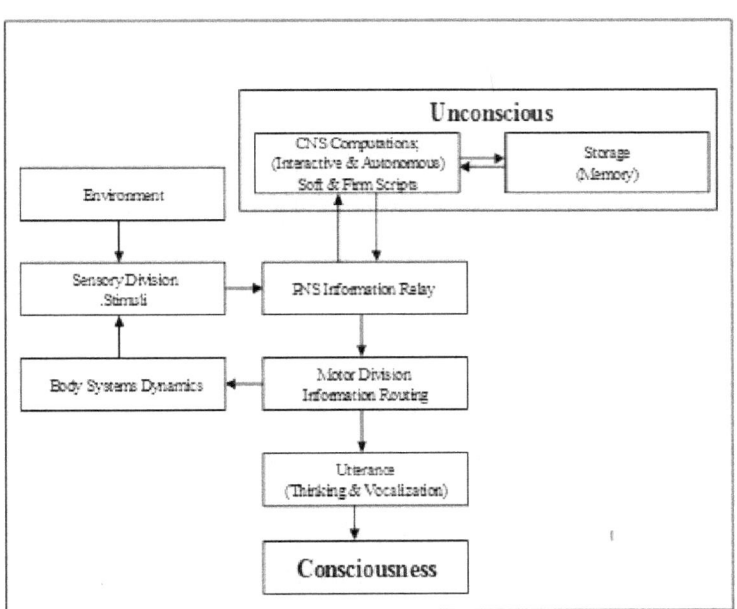

Figure 1- Mammalian system information flow

Philosopher's Stone

● ● ●

I will write about human beings as though I were concerned with lines and plains and solids.

BARUCH SPINOZA (1647)

ALCHEMISTS THROUGHOUT THE AGES HAVE searched for a catalyst to turn base metal into gold; others have searched for an elixir of eternal life. Though the desired ends have not materialized, these efforts rendered fertile grounds for advances in sciences in general, laying out the foundation of chemistry in particular. The aura and semantics of the stone of philosophers has not been lost, and its lure lives on in the human psyche. The mystical interpretations are fulfilling: *"The state of enlightened consciousness;* a universally healing state of being is what is alluded to by the philosopher's stone," according to Frank Metzner.

However, in the realm of philosophy, the stone, inspiring idea it is, may stand for one or more solid premises on which more robust philosophical arguments could be based. Nonetheless, the discipline of philosophy, having gradually overcome the metaphysical mental tendencies of humans treading the waters of our mysterious existence, occupies the first and the highest position among human efforts, focused on making sense of life through inquiries that have always challenged human brains. And it has made headway in many aspects through the establishment of various daughter fields in sciences and humanities, which are the backbones of deep insights and amazing progress in the realm of knowledge. However, its centuries-old stagnation in birthing answers to some of the big questions of life, such as the mind-body problem, reality, consciousness, etc., evinced by the divergence of opinion about them among philosophers, has led to the examination of its reasoning structure by some of today's brightest torchbearers (e.g., Dennet and Chalmers). In such evaluations, the lack of progress is attributed to the weaknesses in the philosophy's argumentative premises. Perhaps one exception, to some degree, would be consciousness: it has been showing better prospects of opinion convergence among philosophers. Actually, it is the subjective aspects of consciousness (called the hard problem), such as feelings, emotions, thoughts, etc., being unexplained, that is the focus of recent heavy philosophical speculations. The behavioral aspects, the easy problem, can be explained with the underlying mechanisms. Materialism, idealism, and dualism have various palatable takes on this issue, but panpsychism, which is finding increasing support among philosophers, holds consciousness as an intrinsic property of matter: a nature of matter, at its very microphysical levels, that somehow builds into animate beings' subjective consciousness. And, of course, if this somehow gets solidly proven, it could be the proverbial "stone" of philosophers—perhaps it could convert some unanswerable questions into answerable ones in the philosophy books!

In this book, from a different perspective, some of the big questions have and are being rigorously tackled as we proceed, and the discussion that follows provides an alternative approach to addressing the question of consciousness. A fundamental explanation of it is offered, and it is based on a reasonable inference from the modern theories of the brain: animate beings' lives are conducted by their central nervous systems' computational operations in response to internal and external conditions, and that consciousness, as argued in the earlier section, is in fact the display of the results of processing and resolutions of complexities that beings encounter in their environment. And this pointedly implies, as difficult as it may be to digest, that all animate beings are but biological machines. Nobel Laureate Sydney Brenner (Woodham 2014) put this overall claim in the proper context during a recorded gathering of scientists:

> *One, how do the genes specify and build a machine that performs the behavior, and two, how does the machine perform the behavior? The answer to the first one is we do not know, but the answer to the second one is that it would depend on the queued memory and boundary condition, like any ready-made machine.*

It is within such context that *we can extend consciousness to everything, animate or otherwise,* since *they all respond to input and display output.* A thermostat displays expansion when subject to heat: its constitutive laws (Fourier's and thermal expansion laws), in tandem with conservation laws, produce the thermal diffusion and expansion—a very simple physics problem. In this light, everything, as beings and things are in themselves (whatever this may imply) devices and react in their own specific way—subjectively, while obeying laws of nature—to stimulants, rendering a display of what may be called consciousness. And so is our consciousness, the experiences of subjective perceptions, which are displayed through our biological interfaces.

While the reasoning for commonality of consciousness is fundamentally sound, animate beings demonstrate an additional characteristic, a distinction from inanimate objects, due to their biological dynamism and modes of display mechanism. However, there are modes of animate consciousness that are similar: deep sleep, upon appearance, emulates such a case. In the meditative practices of yogis, mental registration *only* to the effect of being part of nature, and in unison with it, are reported! Such states of mind are generally developed in abnormal (out of the ordinary) acquiescence of the brain, effected by exposing it to irresolvable, mentally engaging complexities—for example, the Zen Buddhists use koans (Carlisi 2007; Watts 1989), like "what is the sound of one hand clapping," to engender such scenarios. ***The demand on the brain following long periods of persistent, repetitious exposure to nonsensical mantras would engage much of the brain's processing powers—almost all of the neuronal network (likely excepting modalities for biological services), which halts addressing all other demands.*** When brain's backend activities (unconsciousness processes), the ones that continuously drive the external survival needs, are massively hampered or stopped, and the exposed complexities remain unresolved, the efferent (motor) signal are very likely to be muffled—no solution to report—and very weak. That carries very little pulsation of the biologic interfaces, resulting in only minute (animate) mental (consciousness) registration. And in such a state of little mental apperception, experience of nothingness (the loss of ego) could prevail: a case of idling the brain, as far as interactions with the external environment are concerned. It renders almost an inanimate state of being, similar to that of inanimate objects, while unconscious backend processes sustain life and respond to biological demands. ***Rules run animate beings' internal operations, the same way rules run the thermostat—a principle of universal consciousness.***

What we have established here is (1) in the context of input/output systems, everything, animate or inanimate, is ***conscious,*** and (2) that though it is subjective, since what is outputted is what the thing, in itself, exudes, it is still explainable by the objective functioning of the matter's inner characteristics—the physical properties, or biologic machinery, obeying the related governing laws. The subjective consciousness is the display of the subjective output at the biologic interfaces.

The claim of consciousness, being a fundamental characteristic of everything, delineated from a scientific angle here, ironically provides much credence to the panpsychism theory. The context of the computational brain, which served as the basis for the premise of consciousness, could as well provide explanations for other big questions of philosophy. However, that requires alignment with Sydney Brenner's logic that this section started with—a different view of life, which has been substantially but implicitly pursued in this book, all along!

Free Will

● ● ●

No body does evil willingly.

Socrates (500 BC)

There is in the mind no absolute or free will; but mind is determined in willing this or that by a cause which is determined in its turn by another cause, and this by another, and so on to infinity.

Baruch Spinoza (1647)

Mankind has always been coping with many natural or accidental events of various natures, clearly out of their own control, and generally, with little haste, conveniently attributed them to the gods' or God's will and destiny. However, outside of such events, they have always been under the perception of having total control of their conscious minds and of having power on influencing or creating some events and in running their lives. Obviously, this is in accord with the daily practices and experiences of life when, with little thinking and concentration, one seemingly decides and chooses to perform various functions and to conduct actions. Also, the ability to engage in deep thinking and concentration, seemingly by one's own volition, in order to find answers or solutions to difficult problems strongly points to the practice of "free will," which is further evinced by beings' seemingly vast role-play in the theater of life. Many philosophers, up to those in present times, believe in the concept in one form or another. However, in view of the computational brain concept and its ongoing backend operations, it is evident that free will is inevitably a moot subject, and consciousness, despite all its free will implications, is just a mental state, displayed at the brain input/output (I/O) console—perceptions experienced at the utterance interface discussed earlier! Chance, due to the randomness of conditions, is the ultimate determining factor for life and living. And, the ensuing behaviors (mental and physical outcomes) in the process of the interdependent interspecies interactions serve the overall goal of the long-term survival of the species.

In the realms of philosophical thoughts, the role of beings on their life's path has always been subject to very serious study and discourse. Some believe in the ***causality principle*** on logical grounds, though they still allow for some measure of responsibility for one's action, as in the determinist philosophy of compatibilism, while others rely on morality principles that necessitate free will. Francis Bacon equates "will" with "intellect" and believes what reigns in men's minds are "custom, exercise, imitation, emulation, company, friendship, praise, reproof, laws, books, studies, and the agents of social setting, by which the mind is formed and subdued. ***Doesn't all this point to the complete role of antecedent environmental causes in determining the living path?*** However, American philosopher Bergeson, revolting against science-minded empiricists and rationalists who would not heed free will, sarcastically argued that causality implies that the nebula that made our world causes everything.

His logic ignored the fact that every step of this path, from nebula to the present state of the world, is affected by the chaotic boundary condition as well as it is by the antecedent state.

The question has even been addressed by some of today's scientists: astrophysicist Stephen Hawking and physicist Leonard Mlodinow (Hawking and Mlodinow 2010) fundamentally reject the notion of free will, only suggesting it is more pragmatic to allow for choice in one's life. Also, Mlodinow (2011), in a recent book, comes close to attributing almost all behaviors to the subliminal actions of the unconscious.

We note that birth and death are the two most important out-of-control events of the consciousness path, and—aside from birth gene activation and the gene activation/deactivation role in epigenetic—the in-between period of life is also dotted with many occasions of "if I had known better…" and other admittedly out-of-control and life-changing events: this theory connects the dots.

The Burdened Brain

• • •

SURVIVAL, AT ITS MOST BASIC and transparent level, must have practically been the only item on the consciousness agenda for Homo sapiens for the most part of the evolutionary path; it has only been during the last couple of millennia that survival has gradually been strived for in civilized ways—at least seemingly—as far as its implications are concerned in regard to other beings and nature. Among other beings, the drive has remained transparent, the same as ever. In time, skills, inherited traits, and impulses led to the accumulation of knowledge and the domain of consciousness expanded, leading to further sculpting of the neuronal web of the brain. The enhancements of the resolution/solution capabilities of the brain would not be uniform due to irreconcilable factors and inherently would bear variations, which would appear in terms of a spectrum of aptitudes for improved living and advancement among human societies (mainly in the absence of vicious, prolonged, and malevolent subjugations by more powerful others). However, the rate of growth of human mental capacities unequivocally has lagged far behind the rate of the growth of the complexities of life, which are commensurate with progress. The prevalence of modern-era complexities, in the face of deficiency venerability of the mental powers, would possibly render significant adverse influences on the human psyche and conduct of life. And, unfortunately, the societal impacts could be heavy in that the collective behavior is determined by the lowest common denominator among the society members. That is why, despite our immense existing knowledge base, world conflicts and active savagery of humans still persists.

Considering day-to-day life in human societies, contrasted to the lives of our pre–Industrial Age progenitors, our brains are immensely burdened with the pressures of countless problems, which need to be solved! At this stage of human development, the issue of survival, while still rooted in human egocentricity, is compounded by the complexities of modern social living, along with an ever-increasing addiction to the material aspects of life. And this has meant an extra load and demand on the computational machinery of the brain. Therefore, the question of whether the capability/capacity of the brain is up to the par for the timely resolution of all the issues it is exposed to becomes a serious concern. The quality and quantity of the acquired solutions for day-to-day problems would depend on the maturity and sophistication of the individual unconscious, since it is the agency that handles life's problems. The overwork of the brain and/or its exhaustion—problem-solution inability—is the cause of many psychological as well as physical problems. What happens in this case is that unsolved problems keep the brain engaged in the manner of a "hanging job" in computer lingo, and it may go unnoticed or disassociated with, lying under the queue of other posed problems. The effects of unresolved issues in the brain (in the unconscious), while one is not conscious of it anymore due to the onslaught of new problems as time flows, are of a very limiting nature: the full computational power of the brain can get diminished and the severity of the impact of the reduced mental capacity on the human psyche, as well as society, would depend on the nature and complexity of the problems humans face in the conduct of their lives. The inevitable fallouts of burdened brains are most likely the root cause of the troubled psyche of today's humans, the repercussions of which are felt in biological, mental, and social maladies and disorders.

Mental Stress

On the health side, the havoc wreaked by what is called "stress" is quite well known, which we may for the first time clarify as operation mechanisms. Mental stress in humans, on the face of it, is the perception of unease, in various degrees, in response to stressors, such as threats; emotional, economic, socioeconomic, and political instabilities; and other conditions, some due to geopolitical conditions and some emblematic of modern times, which create intense and/or difficult brain engagements. However, since such perception (besides serving as the alarm phase) could also be indicative of other, and often adverse, physiological effects, it has been the subject of much research since the early decades of last century (Selye 1955; Dantzer 1989; de Kloet et al. 2005; Sapolsky et al. 1986; Sapolsky 1990). These efforts, which have resulted in a vast body of very valuable scientific findings, revolve around the physiological processes that *stressors prompt* in the biological system and how they affect its equilibrium, i.e., its homeostasis. But how the physiological processes are triggered through the involvement of the sympathetic nervous system and the HPA axis, as well as other systems—***the underlying cause of what can be considered the overdrive engagements*** of these systems—has seemingly escaped scientific scrutiny and remains unanswered, perhaps because it does not have much bearing on the medical treatments of the ensuing effects of sustained (chronic) stress sensation, and also mainly because many of the details about the brain's processing of sensory data have not been clearly figured out, though the role of the brain in the perception of the world, and life as a whole, has been widely recognized since the Enlightenment era (early seventeenth century). However, progress in the last few decades in understanding the modus operandi of the brain has led to further recognition of the fact that much of the brain's operations are computational in nature. It is inferred that the brain, and the rest of the nervous system, comprise extensive computational machinery that process all sense-relayed information using its innate (evolutionary-formulated) instructions, imbedded in its constructs (as patterns), for analyzing events and creating perceptions. In the context of such idealization of the operation of the brain (Arbib 2009), this work is an attempt to throw some light on the puzzle of stress, suggesting a mechanism for the creation of its sensation and the prompting of the physiologic effects.

Computational neuroscientists, along with a general consensus in all other related fields (as discussed in earlier chapters), are in unison that, one way or another, the brain does calculations, presumably along the lines of what is understood from the operation of today's computers, yet in vastly more powerful and different ways; it is the neuronal network of the brain that owns this advantage. The brain construct–inspired scientific neural networks, which have demonstrated capabilities in solving complex problems with no need for logical formalisms—unlike traditional digital computers, which depend on them—provide only glimpses into the brain's computational system. The extent of its operations, having to do with the (unconscious) maintenance life on the one hand and the (mostly unconscious) management of beings' interactions with the external world (Mlodinow 20011), on the other hand, requiring engagement in resolution and solution (i.e., development of neuronal patterns) of the unimaginable complexities they pose, is indicative of the power and immensity of the computational machinery of the brain and the rest of the nervous system, the understanding of which, despite massive scientific efforts to gain detailed insights into its operation, perhaps to mimic human knowledge and intelligence, will remain mostly, in the foreseeable future, the enigma it has ever been.

However, based on what has been learned initially from the operation of the scientific neural nets, and later from intense scientific research on the neuronal systems and the nature of neurons' data communications—all reviewed earlier—the complexity solution processes of the brain likely (autonomously) involve two steps: first, an inherent implicit resolution (algorithmization) of complexities, and second, a trial-and-error solution approach deploying feedback from the rest of nervous system (Schad 2016) to render perceptions. Upon the brain's exposure to complexity, autonomous attempts at reaching a solution begin, and depending on the intensity of the computational task, the brain can be kept at bay at times, creating what is felt as **stress sensation**, the perception of mental unease. Enduring stress sensation is very likely due to the unavailability of proper existing solution neuronal patterns, or an inadequacy in topological resolution, a preamble to solution achievement, both indicative of continued brain engagement.

Clearly, the power, capacity, and readiness of the deployed gray matter in the processing of complexities—the computational engagement of various neuronal net modalities (subcircuitries)—are much in tune with the "normal" computational demands, defined by the natural environmental settings of species. This has not changed much in humans, in spite of the complexities brought about during the eye-blink period of the emergence of societal life, culture, and civilization, only a couple of millennia long. As such, species' brains in general, having evolved in the natural environment, are configured to handle natural life stressor complexities, and the allostatic load thereof, in order to ensure their biological sustenance. The neuronal pattern (solution) availability allows them to grope along the evolutionary path of the survival of the fittest. A well-known example of this work in nature is in the fight-or-flight response to a physical or life-threatening exposure, which engages the brain heavily, though generally temporarily rendering the perception of unease (that is, stress sensation) and onset of the necessary underlying physiological responses in the biological system accordingly. However, humans, despite their innate disposition to handle the normal complexities of natural life, remain in a very precarious situation since they are mostly unable to resolve the stream of complexities brought about by some of the characteristics of civilized life in the modern era, which burdens their brains over and beyond their natural (evolutionary) computational capacity on a continual basis, limiting brain-processing powers, which can hamper the accomplishment of other brain-regulated biological system tasks as a result. An imbalance or disruption in the operations of the brain explains the perception of unease—an abnormal feeling, the sensation of stress, a natural alarm serving as a sign (a "feeling" biomarker, not to be confused with known biomarkers) indicative of possible physiological effects instigated by stressors.

It is certainly to be noted that all the elements of the biological systems, though subject to impacts of abnormal efferent signals (instructions)—due to stressor-caused disruptions in the brain—most probably have wide-enough safe operations bandwidth to tolerate them, at least for short periods of time. However, its continuation could wreak havoc in the body, possibly adversely affecting its immune system and operations. The overall schema of this thesis is shown in figure 2. In this depiction, senses pick up (stressor) stimuli from internal and/or external sources; whatever complexity they represent is exposed to the neuronal net, and a solution, either final or trial, is issued to the biological system and its interfaces. In the cases of insolvency, due to improper feedback, a nonconvergent operational loop continues in the brain, causing malfunctions in the physiological processes of the systems. It is also in such situations where a sizable portion of the computational capacity of the neuronal net gets greatly vexed, possibly impeding and/or degrading other needed operations; as a result, malfunctions of biological entities, or diminution of mental abilities, can occur. The well-known phenomenon of psychological disassociation is perhaps due to limited capacity of the overwhelmed brain while continued new complexity resolutions and perceptions persist. The proposed schema may have other implications in related fields in sciences and humanities, the discussion of which is beyond the scope of this book.

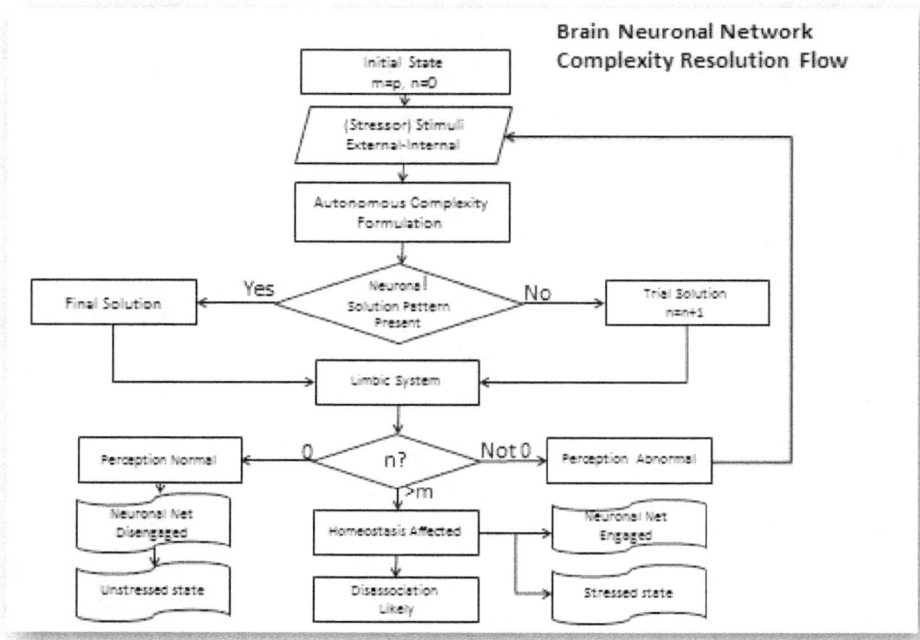

Figure 2. Conceptualization of Brain Computational Flow

To sum up, this thesis suggests that it is the engagement of a being's computational brain with its overall environmental stressors, and the resulting disruption of its normal operations, that triggers the body's improper physiological responses, which are ascribed to stress. The immediate response of the limbic system, which we are capable of feeling, defines the perception of stress. Continued prompting of the rest of the systems under such circumstances instigates physiological processes that ultimately adversely affect the health and well-being of stressed individuals.

The role of mental stress in beings' lives, specifically among humans, has always been recognized. Much understanding of its physiological impacts on the health and well-being of individuals has been gained through rigorous research conducted since the early decades of the last century. However, the mechanisms by which physiological processes are prompted to affect the biological system, as well as the rendering of the stress sensation, have escaped scientific scrutiny, perhaps partly due to the fact that it would not have had any impact on the ***medical addressing*** of problems that are ascribed to stress. On this topic I have followed the plausible theory of stress phenomenon proposed by Schad (2013). The thesis is in the context of the brain's computational complexity resolution and solution apparatus and formulated in a schema, which clarifies the development of the sensation of stress and suggests the triggering mechanism for the onset of the physiological processes, which may ensue its perception. Such understanding supports the long-standing belief in the mind-body relationship, and perhaps resolves one of the big questions of philosophy.

AUTOIMMUNE DISEASES

<u>Immune system disorders</u> *cause abnormally low activity or over activity of the immune system. In cases of immune*

system over activity, the body attacks and damages its own tissues (autoimmune diseases). Immune deficiency diseases decrease the body's ability to fight invaders, causing vulnerability to infections.

In response to an unknown trigger, the immune system may begin producing antibodies that instead of fighting infections, attack the body's own tissues. Treatment for autoimmune diseases generally focuses on reducing immune system activity. (WebMD)

Having explained the mechanism behind stress development, in the following section we take up one of the major diseases of modern times, multiple sclerosis (MS). There is a particular facet of this particular disease called autoimmune disease. The defense system of the body attacks other parts of the body as foreign invaders, an immune system gone haywire! Contrary to the prevalent assumption in medicine, I, following my earlier work (Schad 2015), claim the immune system may be performing at its best, trying to confront and nullify a real impending threat in the system, which would seriously put the system in danger, unbeknownst to the beings; albeit, suffering can result in the process, if the real cause is not alleviated. We believe the real cause is the development of serious measures of stress—rendering certain malfunctions in the brain—which the immune system then aims at preventing their consequences.

We provide evidence for the ***correlation of stressors with afflictions of MS*** throughout the world and put forward arguments in support of the fact that stressors can render ***disruptions in the normal computational processes of the brain***, defining the innards of the ***mental stress,*** which in turn may lead to the onset of physiological adversities in the biological systems, possibly rendering various diseases, such as MS. It can be also deduced that a feeling of unease, normally referred to as "stress" in common usage, is an indicator of a brain-process disruption event.

In case of MS, its link with mental stress, through analyses of various aspects of statistics of ***prevalence*** and ***incidence*** available in the literature (WHO 2008; Izadi et al. 2013; Achiron et al. 2012), can be established. And it would lend itself to the opening up of additional treatment possibilities that could be used separate of, or conjunctively with, the medical approaches. On a grander scale, publicizing the adverse workings of mental stress and its evils may attain statistical gains in incidence reduction.

Physiological debilitating symptoms and abnormal behaviors have been considered neurological disorders since around the fifteenth century (Rolak et al. 2009), and gradual efforts aimed at understanding the phenomena culminated in the 1868 work of Dr. Jean-Martin Charcot (1877). He wrote a complete description of the disease—now known as MS—and the changes in the brain that accompany it, based on what was known at the time and his own clinical studies and observations. Much work has been done since then to find the cause of the disease; some of its intermediary processes have been recognized, and MS is established today as an autoimmune disease, which may also include some characteristics of viral infection as well.

Symptoms such as losing sensation in parts of body, loss of balance, neuritis, tremors, double vision, and urinary problems get the sufferer to the doctor's office, and, generally, an fMRI, blood tests, and neurological tests are conducted. Depending on the results of the tests and the history of the patient, a diagnosis of MS, according to certain criteria and based on a consensus between neurologists, may be made. So far, medical research, in spite of all efforts, has remained in the dark regarding the causality aspects of this disease. It is considered an autoimmune, or immune-mediated disorder, where the brain is targeted, and for which no cure has been found.

The prevalence of MS—the costs of medications and patient care, and the misery of the sufferers throughout the world—has given it the appearance of a very difficult epidemic-type problem, though almost all cases are not believed to be of pathogenic origin. The fact that it affects young people in their most productive period of life has made it a worldwide concern. The prevalence and increasing rates of the disease (incidence) has put it on the WHO agenda, and a comprehensive Atlas report (WHO 2008) demonstrates vividly its seriousness. The following excerpt from pages 14 and 16 of this report, which addresses ***prevalence and incidence*** in total numbers and ***average age of onset and male/female ratios*** throughout the world, speaks to this fact:

Salient Features 1 (affliction statistics):

- Globally, the median estimated **prevalence** of MS is 30 per 100,000 (with a range of 5–80).
- Regionally, the median estimated prevalence of MS is greatest in Europe (80 per 100,000), followed by the Eastern Mediterranean (14.9), the Americas (8.3), the Western Pacific (5), South-East Asia (2.8) and Africa (0.3).
- By income category, the median estimated prevalence of MS is greatest in high-income countries (89 per 100,000), followed by upper-middle (32), lower-middle (10) and low-income countries (0.5).
- The countries reporting the highest estimated prevalence of MS include Hungary (176 per 100,000), Slovenia (150), Germany (149), United States of America (135), Canada (132.5), Czech Republic (130), Norway (125), Denmark (122), Poland (120) and Cyprus (110).
- Globally, the median estimated **incidence** of MS is 2.5 per 100,000 (with a range of 1.1–4).

- Regionally, the median estimated incidence of MS is greatest in Europe (3.8 per 100,000), followed by the Eastern Mediterranean (2), the Americas (1.5), the Western Pacific (0.9) and Africa (0.1). No countries in Southeast Asia provided data.

- By income category, the median estimated incidence of MS is greatest in high-income countries (3.6 per 100000), followed by upper-middle (2.2), lower-middle (1.1) and low-income countries (0.1).

- The countries reporting the highest estimated incidence of MS include Croatia (29 per 100 000), Iceland (10), Hungary (9.8), Slovakia (7.5), Costa Rica (7.5), United Kingdom (6), Lithuania (6), Denmark (5.9), Norway (5.5) and Switzerland (5).

- The **total** estimated number of people diagnosed with MS, reported by the countries that responded, is 1,315,579 (approximately 1.3 million) of whom approximately 630,000 are in Europe, 520,000 in the Americas, 66,000 in the Eastern Mediterranean, 56,000 in the Western Pacific, 31,500 in Southeast Asia, and 11,000 in Africa. The reader should keep in mind that there are no data for some of the mega countries such as the Russian Federation, where the total number of people has been suggested to be quite high in anecdotal reports.

Salient Features 2 (ages of onset and male/female ratios):

- Globally, the inter quartile range for age of onset of MS symptoms is between 25.3 and 31.8 years, with an **average age of onset** of 29.2 years.

- Regionally, the average age of onset is lowest in the Eastern Mediterranean (26.9) followed by similar average age of onset in Europe (29.2), Africa (29.3), the Americas (29.4), and South-East Asia (29.5) and highest in Western Pacific (33.3).

* By income category, the estimated average age of onset is 28.9 years for the low and upper middle-income countries and 29.5 and 29.3 years for high and lower middle-income countries.
* Globally, the median estimated **male/female ratio** is 0.5, or 2 women for every 1 man (with a range of 0.40 to 0.67).
* Regionally, the median estimated male/female ratio is lowest in Europe (0.6), the Eastern Mediterranean (0.55) and the Americas (0.5) and highest in South-East Asia (0.4), Africa (0.33) and the Western Pacific (0.31).
* By income category, the median estimated male/female ratio is same in all income groups of countries (0.50).

The above data contain shocking information about the nature of the distribution of the disease, which can be summarized in the following distinct facts:

1. There is a low age of onset, twenty to thirty-five.
2. There is a similar age of onset among all income groups.
3. There is a low prevalence around the equator, which is mostly less developed as well.
4. There is a high prevalence in industrial countries.
5. There is a high prevalence among higher-income groups.
6. There is a high prevalence among educated groups.
7. The female/male ratio of two, for MS worldwide, increases to three, or more, in less-developed regions.

Also, the following highlights anecdotal extracts from the report, which in one instance draws upon information from an additional source:

1. There is a high prevalence and incidence rate among harsh political and economic conditions in some third-world countries. An example is Iran: the incidence rate has climbed drastically in the thirty years since 1979 revolution, resulting in a change of prevalence from around 10 to around 60 per 100,000, making it close to those of the industrial nations, and the age groups of those affected average of twenty-five years or younger (Izadi et al. 2013).
2. Rates have increased in low-incidence countries with recent economic growth (3).
3. The possible role of one or more combination of environmental factors has been recognized (Rolak et al. 2009).

Added to the panoply of data, we need to also bring in the following medically observed facts about the disease, which speak to its main characteristic:

1. Varying frequency of attacks
2. Varying intensity of attacks
3. Varying nature of symptoms
4. Occasional lifelong dormancy of attacks

Having provided what is mainly known about the distribution of the disease, and some of the well-known facts about the disease, we now proceed to provide the context in which our arguments are pursued.

Immune system attacks in the brain are the medically recognized causes of a number of physiological disorders and diseases. Given this fact, the assumption of occurrences of antecedent events in the brain, serving as triggers for the attacks, is not unreasonable. Since no symptoms seemingly exist prior to the attacks, the a priori events have to be nonperceptible and presumably not involve any measurable physical deterioration of the brain—very likely these are some ***nonphysical disruptions*** in the normal processes of the brain. This, in the context of the present-day understanding of the brain, from the viewpoints in the fields of neurosciences, psychology, cognition, and other related fields, has to be considered a ***disruption in its normal computational processes***.

The correlations of MS with some of the seemingly advantageous elements of life in the industrial and modern era are indicative of the (mostly unconscious) heavy demand on the minds, which such conditions inherently beget. Likewise, the correlations of harsh societal, economical, and environmental conditions—generally politically induced in today's world—with the prevalence of the disease again point at the prevailing heavy mental engagement as a result of copying with these adverse life-affecting elements. It is therefore clear that the high ***incidence*** and ***prevalence*** of the disease correlate with the excessive engagements of the mind with the irresolvable complexities posed by modern-life environmental stressors, possibly rendering disruptions in the brain's normal operations. In the context of a computational model for the (mental) functional operations of the brain, any nonphysical disruptions in the brain's processes immediately indicate its engagement with ***irresolvable computational problems,*** which can adversely bear on the physiological processes, which are computationally controlled by the brain. An early indication of this dilemma is perceptions (feelings) of measures of unease, commonly referred to as ***stress***.

Stressors have been a fact of life in any era of human civilization; however, since stressors of different natures and magnitudes have been added in the post industrialization era, their nature and implications would be different from those of the past. With the advent of the industrial age, progress, as well as problems, began affecting human life very dramatically, mostly for the better and at times for the worse. One of the major adverse characteristics of the new era is the presence of high, "sustained" stressor levels—though mostly directly unnoticed—a condition that very likely did not develop in the prior simpler life on the planet. Obviously, human beings suffered much from many hardships in the preindustrialization era. However, those mostly related to the issues of food and shelter; the adverse attributes of thoughts and desires of material ownership and other worldly accolades, since mostly limited, did not make significant contributions. Of course, wars and pestilence (not affecting major segments of humanity today) would bring occasional very high peaks of stressors. But the resulting stresses, whether they were long-lasting or not, were handled accordingly through the mental agilities gained in the process of natural living. Today's high levels of (background) "sustained" mental stresses—mostly not (natural) survival-driven—are brought about by new stressors: the challenges of modern life and goals of high professional and material achievements, as well as by the (mentally) demanding efforts of chasing the (mainly) illusory premises of materialistic or fantasy-based happiness in the postindustrial world. Drastic changes of habitat, modern lifestyles, and increasing family instability, along with the many synthetic compounds in the environment, all mostly unfamiliar to human biology and psyche, are also significant contributors to high stress levels, as well as the causes of health problems, though some of the former can also be consequents of the latter.

Stressors are discerned as complex problems in the brain, and when they are irresolvable, their effects, based on scientific inference from the present understanding of brain functions, must be hampering the proper deployment of neural patterns (brain web constructs of organized and synchronous neural signal firing schemes) that control most inner body functions and the body's activities. This may drastically affect hormonal balances as well as other bodily functions, upsetting homeostasis, the equilibrium mainly regulated by brain signals received through the nervous system, engaging the endocrine system, the autonomic nervous system, and the immune system as needed. The state of the brain that leads to any measure of disequilibria of the biological system is a ***stressed state*** and ***its immediate effects on parts of the limbic system may create abnormal sensations, perceptions of which are what commonly is referred to as stress***, as indicated earlier. The survival of species has obviously depended on certain level of stress, that is, stressor-originated effects for certain necessary physiological and bodily responses, such as fight-or-flight and other survival benefits; however, sustained levels of high stress (continued disruption of the brain's functional processes), in view of links with the rest of the biological system, may initiate upheavals in the body, forcing the immune system to rise to the occasion, as is the case in MS, attacking whatever and wherever the source of the problem is. In this case, it attacks the brain to eradicate the cause, ***a response expected from a "strong" defense system***! Another reason for an attack of the immune system would be the presence of infectious viruses that can act as a trigger; however, since the prolonged presence of viruses is not likely, this scenario does not conform with the observed long-term persistence of MS. Perhaps whatever short-term neurological effects emanate from the viral infections can be put in the category of known MS afflictions that have been found to be of very limited consequence in life. In MS, the attack removes myelin from a number of nerve cells (affecting the quality of the neural signals, at least temporarily), which may get repaired, perhaps immediately. Mild attacks affecting small areas of the brain may go unnoticed. Repeated autoimmune attacks leave some cells with lesions. If such attacks continue more severely, and/or for a long time, more parts of the brain may be affected and very dramatic events in the body may result. Field evidence suggests these attacks may, in some cases, stop at some point and never happen again; however, in other cases there are stop-and-go episodes leading to various stages in the progression of the disease.

Considering the high prevalence and incidence of MS in the stable, advanced industrial countries, as well as in the rapidly progressing (developing) countries, and in parts of the developing world with socioeconomic and political instability (affecting women much more than in the former cases), the commensurate stresses, the obvious indicators of quality of life in such high-stressor environments, would very likely play a major role in the onset of the disease, indicative of a strong correlation with both modernity and modern misery (the latter applies worldwide).

The two exceptions of the industrial countries, the Netherlands and Israel, which show a much smaller prevalence, can perhaps be explained by the presence of very strong social welfare systems and cultural ties in the populations—both play a major role in stress reduction.

The lower statistics in less-developed societies, though varied, may be attributed to the prevalence of lower-strength (or at best normal) immune systems evidenced by the occurrences of all kinds of bodily malfunctions among their population. Somewhat of anecdotal evidence may be in the fact that people more affected by infectious diseases (possibly due to the lower strength of their immune systems) are less likely to get MS. Of course, lesser pressures of modern life would be an added factor, as well.

DISPELLING THE MYTH

Last but not least, we need to dispel the myth of autoimmune disorder. Underperformance of the immune system—that is, not providing a proper defense for the body—is quite understandable on the grounds of its weakness, which could have genetic or environmental reasons. However, immune system attacks on certain functions of the living system, while protecting others, poses a dilemma that is hard to resolve, and that is why it is simply called a disorder. As noted in the above quote, the apparent adversity is brought about by unknown triggers. ***Considering the immune system an evolutionary mechanism of defense, its function—though medically misinterpreted and contradictory—must be to stop the triggers,*** and ***it is the collateral damage that is observed as the autoimmune disease.***

The seeming attack of the immune system on the self is to eliminate the real causes of the malfunction of the elements of the body itself, which is wrongly interpreted as its failure. The case of MS is very telling—this disease is caused by wrong signals from the brain, which may drastically disturb the body's homeostasis. The immune system's proper action is to stop the signals; it stops them by creating lesions (affecting the myelin sheath of axons). Obviously, if the wrong signals continue, the development of many lesions renders the body incapacitated and what is known of the disease comes about. The remedy is to eliminate the real cause, which in the case of MS, is the proper handling of stressors!

CLINICAL EVIDENCE:

Stress management (Mohr 2012) has already proven its effectiveness in the dramatic reduction of attacks in patients; it serves as an initial step in the path to confirming the hypothesis. Important evidence that has a favorable bearing on the hypothesis is the widely accepted curative role of the placebo effect: its efficacy has been proven in many clinical trials where the hopes of a cure fight off the effects of certain stressors. The validity of the thesis can be investigated in the likes of the following suggested studies:

1. The development and implementation of a vigorous stress-management program in groups of patients whose attacks happened during a highly stressed period of their lives and the monitoring of the results.
2. The evaluation of episodic attacks in patients to find out if they coincide with high-stress periods and monitoring the rate of the creeping cell deterioration during the quiet periods.
3. An investigation into the conditions of the lives of all once-afflicted people who lived a normal life to old age, finding out when and where they were afflicted in order to find the needed correlation (reportedly some 20 percent of diagnosed patients have experienced such good fortune).
4. Collecting and analyzing fMRI results unrelated to MS in elderly people, addressing any presence of lesions in people who have no MS symptoms.

It follows directly or by inference from our hypothesized genesis of the disease that in the development of brain lesions, the intensity of the stress and the strength of the autoimmune system, as well as the levels of the agility of the minds and mental attitude, matter significantly. Brain agility, problem (stress) resolution ability, lifestyle, and having positive outlooks in life are all important factors in avoiding the disruptive actions of stressors. Only abnormal, out-of-control circumstances may be overwhelming enough to induce serious disruption in brain processing functions.

Considering the episodic action of the autoimmune system in MS, it is plausible that after an attack, relief from all kinds of hanging (sustained) stressor-posed problems happens (an electric reset of sorts), excepting those of relevant ordinary concerns. In other words, the stress-related neural patterns are modified and/ or destroyed, and erratic signals are stopped (neural pattern signal weights need to be reconfigured depending on the subsequent conditions of the person's life). Perhaps that leaves the brain with a much cleaner slate, at least for a while, and hence no need for the continuation of attacks—the enemy is stopped, at least temporarily. In some afflicted persons (the lucky ones), the levels of preattack stress are never reached again. The postattack condition of the patient's life may perhaps be conducive to permanent, automatic removal of high stressors, and therefore stress—and 100 percent normal life continues. And, in some other cases, lifestyle changes and/or other natural or medical measures may either prevent the continuation of that level of stress or confuse the immune system, preventing future attacks, and allowing a normal life to go on. Deploying the plasticity of the brain to enhance its data processing and complexity resolution capability through learning and engagement with arts and sciences and learning new skills would serve as a tremendous added measure to help a postsyndrome normal life.

To wrap up, I suggest that stressor-induced mental stress is the reason for the observed direct correlation found between environmental stressors and the onset of MS. Stressors stem from the adverse aspects of modern life, which are discerned as irresolvable complexities in the brain, hampering its normal computational processes that control the biological equilibrium of the physiological system. This hypothesis finds support in stress management studies and also indirectly from medicine's acknowledged curative role of the placebo, which establishes the mind's role in disease treatment. As such, the role of stress in any disease, compared to that of pathogenic factors, should be given due weight in diagnoses, as well as in prognoses. This view of scrutinizing life under much higher "sustained" levels of stress—a circumstance that the evolutionary path did not prepare us for—carries much substance.

Hallmarks of Hereditary and Evolutionary Knowledge

● ● ●

BEFORE ADDRESSING VARIOUS BRAIN PHENOMENA of interest to this section, and discussing the possible root commonality among them, an important point discussed in the early part of the book needs to be brought up again. It has to do with the nature of brain computations, which makes it relevant to discussion here. It is the fact that parallel processing at the heart of the unconscious (backend) computations of the brain—of tremendous advantage in data and information handling, compared to the serial approach—is what makes the upkeep of the massive tasks and handling of various phenomena possible. In this relation, the prospect of quantum computations, having the advantage of lightning computation speed—hinted at by the likes of British mathematician and physicist Roger Penrose—may not sound too far-fetched in view of the instantaneous mapping of objects such as a face, involving millions of bits of information into its perception upon conscious demand. With this in the background, we continue.

A host of phenomena, such as sleep, dream, hallucinations, hypnosis, burdened psyche, and psychosis, has been the subject of a great many scientific studies aimed at understanding their underlying mechanisms. The scientific inquiries may also offer possibilities of glimpses into some of the secrets that the brain may be holding.

What may make this possible is the fact that these phenomena are possibly the results of the interplay of the consciousness and the unconscious, in the semiaware (of various degrees) state of the mind, which may be called a subconscious state, as part of the ever-occurring brain activities. And if true, this serves as a further validation of the inner workings of the computational brain.

Sleep initially was thought to occur because of the weariness of the body as a result of conscious activities; however, it is now known that it happens partly due to the release of pituitary gland hormones in order to restore brain tissues and consolidate memories. Additionally, sleep may also be partly due to the mind's response to the stimulus of darkness, learned from early man's experiences, when hiding in a corner and staying quiet from possible hazards was at the top of the totem pole of survival.

In the computational brain, sleep may also be due to the temporary input/output halt condition of the brain computer, with all the capacity being at the service of both physiological and mental needs. This is the time at which no new input of problems and data can be taken in, and hence the interactive processes of the consciousness have to shut down. In such a scenario, as repair and maintenance of the system rapidly progresses, the freed up computer (metaphoric CPU) allows the queued in (leftover and unaddressed) demands in the subconscious to be addressed. The download of outputs during sleep renders dreams, which are mostly nonsensical due to the absence of consciousness feedback. As soon as running of the dreams starts taxing the performance of the maintenance work, they are stopped and another cycle of sleep starts. The gradual lightness of sleep and the increasing length of REM (rapid eye movement) sleep are indicative of less demand on the computer and queued task execution for further streaming of the unattended (unsettled) perception experiences. Finally, upon completing the settlement of the memories of the events and the needed maintenance, the brain computer becomes available for new input and complete waking and consciousness begins.

Sleep, as a well-studied phenomenon, is proven to have generally five cyclic stages, all of which involve some measure of consciousness, throughout which the brain remains active. Each cycle lasts for about ninety minutes, with increased levels of awareness, and the REM sleep that appears at the end of each cycle gets longer as sleep comes to an end. In the first stage, one may experience fantastic imagery and various sensations, such as falling or flying, and at the following stages—the REM sleep—the dreams occur.

Many plausible explanations for dreams have been provided, the most viable of which are the following (Myers 2001):

1. Freud's manifest content postulate—creating a storyline that is gratification of a censored version of our latent wishes;
2. Synthesizing neural bursts into a storyline;
3. Helping process information from the day and fixing it in memory;
4. Serving physiological functions.

These postulates are very different yet they all are correct, in the sense of a blind group of people trying to explain an elephant. The gratification of some wishes in dreams is a common experience. Many of the complete and sensible dreams, with clean-cut scenarios and vividness, fall in the category of Freud's explanations: the experience of controlled dreaming, by focused contemplations before sleep, may attest to this fact. The second postulate truly relates to the onset of the random firing of signals in the neural net originating from the biological activities of the body. Bits and pieces of unfinished daily ponderings, being attended to when conscious activities are low, account for the third category. The fourth explanation relates to cognition development through the plasticity property of the brain in response to queued learning during conscious periods. Long REM sleep periods in infancy correlate perfectly with such expansion of the neural web of the brain. The "slide show" type dreams of children, with little or no storyline, are indicative of signal firings in the web as a result of new synapse connections. Two of the explanations relate dreams to consciousness data, and the other two find the source in biological and physiological events. However, there are dreams with no hint of consciousness, consisting of fascinating scenarios that are bound to have originated from the unconscious. Such are like hallucinations, which are unreal sensations that one may experience, and these often occur in the first stage of sleep. They may relate to the hormonal stimulation of neural nerves, rendering an avalanche of firing in the neural web, the likely result of which could be imagery or sensation.

In dreams, it is the submission of queued thoughts to the brain (inputs to its computer likeness) that renders all real-seeming felt sensations and experiences when no external visual stimuli are present. The vivid visualizations that happen in dreams must be the results of the mind's simulation work, which undoubtedly uses its bundles of (burned in) related phenomena formulations—the natural governing laws—to produce the effects. The rapid eye movement that occurs during dreams, suggested by some researchers to be a process helping the oxygenation of the cornea, may also be somatic phenomena of the simulation of the dreams in the brain, i.e., related to "seeing" the dream. This can be better understood in the context of the seeing phenomena. In the process of normal conscious seeing, the object's modulated light reaches the sensory nerves of the eyes and a model of the object is formed in the brain. Simulations, as explained in the earlier section, are the major function of the brain in developing a representation of the physical world in which life is sustained. Bats use sound waves for the simulations they need. Today's technologies of imaging hidden objects, such as ultrasound or penetrating Radar, similarly use waves to define the unseen. The knowledge that a very sophisticated mathematical formulation of the wave propagation phenomenon—at the heart of such systems— is the key to them working, supports some of the fundamentals of the workings of the brain discussed in this book.

Hypnosis is an induced state of consciousness in which the subject's voluntary power of action is lost. At this state of semiquiet brain, the subject's increased susceptibility makes suggestions or inquiries more effective. Vulnerability to the planting of false memories is of serious concern here, and it is essential that subjects are warned a priori to the inducement of a hypnotic state. Hypnotherapy is a controversial method for behavior modification.

The consciousness is also burdened with all the encounters of daily life: problems, events, and even thoughts are categorized by the pain-pleasure polarity of early perception and continue to engage the brain for solvency. Some are immediately or timely resolved, and there are those that will remain unresolved. While the conscious demand for resolution may get buried under new mental engagements, or get psychologically disassociated with consciously, the process for resolution continues in the unconscious. A lack of resolution is generally due to the insufficiency of proper data or the inadequacy of the computational power of the brain, due to the accumulation of other unresolved issues. A high-intensity pain category problem, such as an infantile trauma, that remains unresolved may form a psychological complex, becoming part of the landscape of the hidden storehouse of the mind.

Regardless of the causes of the states of out-of-the-ordinary brain engagements, the vast spectrum of data—dealt with one way or another in the unconscious brain—is "mind boggling." Unresolved psychological complexes, grotesque figures of childhood dreams to the experience of archaic figures, audio-visual sensations of music and real-life vividness of other dreams, and sudden occurrences of (forgotten) sought-for information, along with other facets—touched upon throughout the earlier chapters—point to a treasure trove of information in the brain. Some of the data relate to conscious experiences, as discussed earlier, while others must originate from the depths of the unconscious, perhaps formed in ancestry or in the evolutionary process.

As is known, dreams have already given clues to a host of elements and events in the unconscious or subconscious, which have been described by various schools of thought in psychology. Freud's reason for dreams is the basis for clinical psychology treatment, where hypnosis is also occasionally used. The reason for the effectiveness of this approach to treatment can be easily explained from the perspective of the computer system model of the brain: when repressed or forgotten unresolved pain category problems are revisited later, with hypnosis or otherwise, there is the possible exposure to the benefit of a more expansive consciousness (i.e., better data and a more developed neural web and perhaps a better formulation of the problem, which may be the case for many patients). As a result, a successful automatic resubmission occurs, as it is normal to any thought process, rendering resolution and the consequent freeing of the load of a hanging job from the processors of the brain. This is why such treatments, repeated over a number of sessions, have a fair chance of relieving patients' psychological ailments.

As it stands, the storage element of the brain is the depository of many of the lifespan's experiences, and there is general agreement to the inheritability of some of the behavioral patterns and traits. However, this may not be all; there is a strong argument and evidence in favor of the idea that the blueprint of life holds much more in addition to the biological secrets of its evolutionary past.

Psychosis, and a host of other serious mental disorders, is a state of the brain in which the sufferer loses touch with reality and experiences a totally altered perception involving all the senses. The question of where such discernments originate from, and whether such psyche phenomena are offering a window into the secrets of our collective unconscious, are what disciples of the Jungian school of psychology endeavor to answer. Carl Jung, the founding father of this school of psychology, believes that "the human psyche is as little personal as the body": "It is rather an inherited and universal affair." It is believed that he bases his famous brand of psychology on the nature of mental imagery from schizophrenic patients—he may have been one at some point in his life; his recently published drawings may relate to this episode in his life. These images exhibit peculiar mythological, legendary, or generally archaic characters. This take on some aspects of the workings of the brain leads to the idea that the unconscious database of the brain may include some of the past collective consciousness, which is handed down genetically. This claim has fascinating and outrageous implications. Be it as it may, one should note that the rich, and often similar, mythology and religions of the world, which have much commonality, seem to have evolved around the imagery experienced by the tribal dreamers and prophets in various cultures.

One may assume that the reported imagery is the mind's fabrication of what may have been experienced by mankind through the ages. However, it would be more plausible if one assumes that these effects have roots in the early stages of the development of what became consciousness of sentient beings, when the ongoing fireworks (bursts) of the neurons left their imprints in the evolving brain constructs (patterns). These events and imageries evolved through hundreds of thousands of years of man's reflections until they developed into the mythologies of historical times. The imprint of such developments in the form of genetic predispositions, i.e., making them inheritable, provides support for Jung's interpretation. It should be emphasized that such plausibility is cautiously ascribed to phenomena that evolved over millions of years, though this is still minute in the scale of biological evolution.

Relying on the premise that the topology of the web of the brain is the essence of the brain's neuronal structure and patterns, and that it is shaped, reshaped, and configured in response to the calls of natural life, it is as well plausible that some information relating to the experiences of conscious lives along the path of evolution is also woven into this fabric of neurons as instructions. And that this, in the face of the formation of the blueprint of life with all its magnanimous intricacies, developed in the evolutionary process does not seem too speculative. If true, a glance at the roadmap of the evolution of beings makes one wonder what imprint each premutation stage of life—each of its animal forms and its environment—has left in the web of the brain. Perhaps what in the case of mankind is called man's "dark side," was left, riding on the impulses of primeval survival instinct. The acts of savagery among humans—wars, torture, etc.—rife in various misguided rationalism, which still strongly persist today, must relate to such a trait!

The results of controlled drug-induced experiments may add some weight to the above speculation. Perhaps it would be stretching the above premises too far to refer to the beautiful figments of imaginations or the hallucinations of Carlos Castaneda. Along the same lines, tribal ceremonial (natural) drug use can be mentioned. Some experienced perceptions by such practitioners have been written about and made into movies. Whatever argument one may use to discount the reported phenomena, it is impossible to convince the experimenter of what has been experienced. One may raise the argument that the strong inhibition-neutralizing effect of the drugs may be what is needed to establish a communication channel to the unconscious and transcend what is defined as normalcy in the mind. After all, if the mesmerizing drumbeats of the peyote-smoking tribal party allow the White Eagle (aborigine) to soar in the sky, the way eagles do, one is bound to admit that something fantastic has been experienced, adding substance to the speculations. It is important to note that though all such events are in the final analysis mere simulations by the brain, the data for the simulator are the kind that the consciousness is not likely to be able to provide. Apparently, similar sensation are felt by some who experiment with LSD; feeling detached from the body and able to fly, in some cases, has had very unfortunate results. However, it is important to mention at this juncture that legal controlled experiments with this family of drugs, one reported recently on NPR's LabRadio (www.labradio.com), have shown fascinating results: in this trial one subjects quit long habit of smoking and one became religious, which are noteworthy because of the brain reprogramability connotations; some subjects experience a hard-to-explain sensation of unity with nature. (This was also addressed in the earlier sections.)

The powerful controlling force of religion, which starts with voluntary submission, has its seed in a common collective consciousness in the form of awe and fear of natural forces that bred helplessness in humans. This, acting as an internal stimulus, creates the vulnerability that resulted in the evolvement of the power of religion and the need for it. From Plato on, many thinkers who proposed brilliant schemes for the advancement of human societies felt the need to add a religious mouthpiece to it. Major religions basked their ideas in parables, metaphors, and allegories, which connected with the predispositions of popular minds. Spinoza is perhaps right:

The people will always demand a religion phrased in imagery and haloed with the supernatural; if one such form of faith is destroyed they will create another.

As mentioned before, there seems to be no personal aspect to some of the abnormal sensations, and explaining them in the context of our computational perception model, as we've done all along, required the use of arguments and reasoning that in our final analyses may not have been very solid. However, in the context of the overall model of the mind pursued in this book, a consistent theory explaining the knowledge potential of the brain has been set forth (further discussed in the following section).

Plato's Knowledge World: The Brain Web

● ● ●

What appears to modern man as pre-scientific mysticism can, from a broad view, be understood to be a religio-scientific expression of intuited order in the working of the universe.

N. Stiskin

Let us enter within, if we can fair out the ultimate nature of our mind, we shall perhaps have the key to the external world.

Arthur Schopenhauer

THE CREATIVITY OF HUMANS, THROUGHOUT the history of civilization, has been the engine of the gradual collective human mental development and progression of the conditions of their lives. Great artists, ancient mystics, philosophers, and scientists of all brands make the list of geniuses to whom humanity owes much. Getting over the mesmerizing aspects of their creations, the curious mind cannot but ask what the mental phenomena behind their accomplishments are. What renders such creativities? A talent for learning and training, while serving as a stepping stone cannot, under sharp scrutiny, explain the phenomenon. Then how does it happen; is there a source of knowledge that geniuses draw upon? In the following section, we will make an attempt to come up with an answer to this inquiry, which has always been on curious minds.

Two categories of evidence make the existence of a world (source) of knowledge—beyond the grasp of normal minds—more of a plausible reality:

1. The first relies on certain human creativities and discoveries that defy any (personal) consciousness origins: the wonderful musical compositions of the likes of Bach and Mozart; the fascinating, mind-boggling art, like that of Escher; the great theories like the theory of thermal energy as the building block of the universe, originally offered by pre-Socratic philosophers; the very early knowledge of evolution evidenced in the historic literature of different cultures (e.g., Plato and Lucretius, 99 BC–55 BC, which left Darwin to prove it); the power of nuclear energy—a captive sun inside an atom; the poetry of the Persian mystic Rumi; the hints of deep knowledge among philosophers, glimpses of which have been presented in this book; and finally the fascinating scientific discoveries of the past century, continuing to present day.
2. The second is based on the amazing reported imageries experienced by schizophrenic patients and the occasional strange dream experiences of others; the rich and similar mythologies and legends of different cultures; and more.

When the processes of these very out-of-the-ordinary events are examined, the necessity of a source for them can easily be argued for; in the face of the unavailability of the notion of a computational brain until recent decades, the supposition of the existence of a world of knowledge, external to our world, would be very rational. Initially, Plato suggested it, and many still justifiably subscribe to it. A favorite anecdotal support for this thesis is the phrase *"it comes to me,"* which some great minds offer in response to the question of how their wondrous creation or discovery was accomplished,

Given the advent of the understanding of the brain and the recognition of the nature of its computational operations, it should be apparent that the treasure of knowledge must lie within the self, only accessible to seekers with the right map. To substantiate this claim, we need to again review the computational essence of the brain in light of today's knowledge about digital, neural, and perhaps even quantum computers and computations.

When comparing the familiar digital and neural network computational operation and processing systems—both solve problems—one's attention is bound to focus on the underlying difference between them. However, before attending to this issue, it is beneficial to briefly review again the advent of traditional computers in the realm of their use in the development of artificial intelligence. Obviously, we are also cognizant of the fact that almost all recent scientific progress is owed to a major degree to the availability of digital computers. Though the initial force behind the effort to mimic human intelligence—based on the use of symbolic language programming of the thinking process—has been diffused to a great degree due to situations of absence of logical structure in some aspects of thoughts, such as intuition, nonetheless the achievements have been impressive. Many behaviors and thoughts have a logical structure to them and can be duplicated through programming. In traditional computers, the use of algorithmized instructions, written in a symbolic language, is what makes data processing and problem solving possible. Basic rules and laws, expressed in mathematical language, have to be turned into computer language (**resolved** as computer codes) before the task can be submitted for a **solution**. However, neural nets, inspired by the structure of the brain, work on the basis of training: knowing the solution of a problem (submitted to the neural network) a priori, the conditions of the network nodes, where many mixed signals of the **input** are received, are manipulated until a match between the **output** and the expected results is achieved. After such training, new problems of the same class can easily be solved. As noted earlier in this book, no programming is involved. The fact that a problem can be solved in this manner should leave no doubt that some embedding of basic laws, or the rules, governing the problems takes place in the (scientific) neural web (nodal patterns, or structure), creating its imminent resolution and solution capabilities. It is such underlying properties in the neural network that allow for the handling of complexities for which governing laws are not known or difficult to formulate.

We have reasoned before that based on three fundamental congruities of the neural network with the presumed brain computer (neuronal network), namely (1) the general web (brain inspired) structure, (2) the trial-and-error learning approach, and (3) the ability to mimic some measure of human intelligence, we can infer that the brain is very likely to operate, in the least, in the general manner of neural networks and certainly with power and attributes that surpass any imaginable progress sciences can make in the way of improving them. Of course, much evidence from neurosciences research serves also as a solid basis for the extrapolation. And, since complex life has evolved while being affected by natural forces, fields that govern all natural phenomena, its central nervous system with the brain at its helm, having had the task of its sustenance both internally and externally—in harmony with cell-level machinery developments throughout the biological system—had to unequivocally be endowed with a vast brain library of implicitly embedded natural laws. And this must have been scripted in the scroll of life, the DNA, and in it, the scheme and precepts for brain structure.

DNA is a double helix strand of certain molecules, which has been gradually organized from the onset of life on the planet. It dictates, governs, and provides for the autonomous upkeep of the species and the transfer of the imprints of their adaptations and learned skills to their progeny. This dynamic recipe of life renders varied forms and brands of beings (through gene mutations), the continuation of which has been at the mercy of natural selection in the chaos of the life environment. In surviving beings, the biological forms and the structure of the central nervous systems are commensurate with the natural sustenance needs of each species. In humans, nature has been at its best to provide for the most sophisticated structure, the results of which have put humans on the path to civilized living. The human brain is a massive web of neurons scripted for the resolution of all complexities—whatever law they obey—in the implicit form of ready-made equations amenable for trial-and-error learning solutions. Additionally, DNA and its dynamic gene expressions would also engender brain web embeddings reflective of prolonged life-affecting experiences in various stages of the evolutionary past—dramatic and traumatic and otherwise—which were continually registered in the brain constructs of many generations, perhaps in the form of latent memories in the neuronal net of the brains of the successors. All phenomena that affected the evolving topology (patterns) of the brain must have become part of the genetic predisposition of beings—the role-play of adaptation, in the way that survival of the fittest is the progenitor of such genetic imprints—to keep the survival machinery churning successfully. Obviously, this is a very speculative statement, but the concept of the ***collective consciousness*** phenomenon, suggested by Jung, adds some credence to it.

We sum up this part of the section by a quote from Mlodinow's book:

An inward light…a light without which the human race would long ago have been extirpated for its utter incapacity in the struggles for existence.

(Philosopher and scientist Sanders Pierce)

Now, with a machine such as the brain, laden (implicitly) with the most sophisticated formulation for every law of nature, even perhaps the workings of the universe, the potential for discovery is so great that it is beyond rational thinking. ***The brain is likely to be truly an encyclopedia of the abstract working models of nature and a depository of knowledge.*** Life itself is a testimonial to the workings of such a knowledge system: DNA program scripts, through the expression of the genes it holds in the machinery of the cells in general and in the renderings of the brain constructs in particular, are running to keep life going, embodying the mathematical expression of fundamental governing physical laws.

Had a traditional computer equipped with such firmware (code) capabilities—emulating the governing laws of physics—been available, all one would need to do to get any phenomena simulated would be writing the proper script, calling out the appropriate codes and data, and running it for computation. To access the resources of the brain, problems need to be defined and formulated (mentally scripted): this approach is what is practiced in the arena of research by trained professionals, who work their brains for long periods to formulate their problems of interest, and at some point the solution happens. ***It comes to them***, ***downloaded as their perceptions***, the "Aha!" of the particular enlightenment they were searching for.

It seems like when the right question is formulated in the brain (the right thought), the answer eventually appears.

However, the shaping of right question itself entails an evolutionary process. In his recent book, Professor John Kounios of Drexel University subscribes to the validity of the "Aha!" moment as an indication of insight and achieving solutions for new problems without a set path, as is practiced in scientific research. However, he misses the fact that path or no path, every solution is preceded by an Aha! moment (with or without the pronouncement), in the context of the computational theory of the brain.

In the scientific approach to discovery, the process entails (1) narrowing down the subjects of inquiry, which allows for intense focus, and (2) using mathematical expressions that are naturally in line with the brain's innate, implicit mathematical resolution language. Of course, all discoveries proceed in steps, following the many "Aha!" episodes that guide them. Thus far, the uses of symbolic representation of phenomena, their mathematical expression of the governing laws—the ones already discovered from the book of the brain, which formulate and align abstract models, the way a trained mind understands (this is achieved after expansive corrective efforts)—have rendered further solutions and discoveries that humanity has much benefited from. It should be noted that the evolvement of the sciences has taken the collective efforts of mankind over a couple of millennia, from which one may deduce that disciplined efforts by scientists, in time, will gradually unlock the secrets of much of the universe, as more focused and correct questions get posed. Therefore, if humans do not succumb to the misfortunes of collective ignorance, represented in ignorant power structures, the mysteries of the universe will gradually become more and more understood, and the *sky is not the limit*! However, revelations about the brain's knowledge base, as evinced by the successes of science's systematic approach (within the confines of specific disciplines) do not negate other possibilities for its access. Discoveries have also happened in the focused corner of genuine mystical practices. Considering the processes involved in the disciplined efforts in scientific gains, one cannot but wonder, how did discoveries in the pre-Renaissance era, when no major institutions and no system were in place, come about? Certainly the process must have been different for the earlier investigators. From what is known about the beneficiaries of the occasional revelation in the past, and from the theory of the workings of the brain laid out in this book, the ability of the early man and the mystic must have relied on certain premises for tapping into the treasury of the knowledge of the unconscious: a tranquil brain and a very sharp focus on the subject, a trance-like, meditative state of the mind, prevailing the psyche. The conditions of materially unengaged, focused thought—that is, fixing one's gaze at the natural phenomena and being bedazzled by it, rendering the right question. *In the realm of the brain's computational design, mental formulations (scripts) of problems are written in high-level language of thought, which are then converted to the machine language of the brain for the solution!* Such scripting may be what was behind the early revelations. Also, what possibly helped the needed formulation of the problems and their subsequent solutions (the perception experience) may have been due to their thinking in shapes (geometry) rather than words, as some later great thinkers believe (e.g., Spinoza). Finally, from another perspective, with beings always exposed to some of the fireworks of natural phenomena, it is not hard to think of a common trait of experiences on life's path that perhaps became the source of the myths, legends, and archaic imageries that are embedded in the collective unconscious.

Some perceptions seen in very high meditative states are inherently ill described due to the inability of language. *Perhaps when possibly the score of the symphony of nature, of which beings are a part, gets played in full*, it is not explainable and can only be heard when one is in it! Schopenhauer says, "Thoughts die the moment they are embodied in words." There have been efforts by genuine mystics of different cultures to symbolically reveal their experiences or guide other seekers to attain such enlightenment. The twelfth-century Persian mystic Shorevardi writes about his experiences in such symbolic ways; such efforts must serve as sources of the many mythologies and folklore that provide the basis of some religions or life ideologies. A glance at the Bhagavad-Gita is much telling.

The shortcomings of language are easily noticed in efforts trying to explain the sensation of touch, or the sound of a sliding leaf on a cement sidewalk; it's the same in the case of many other phenomena, around which revolves much philosophical argument. The only way out of such unending discourse is to agree that not all experiences, mystical or otherwise, are amenable to logic or explanation. *But perhaps for the ultimate meditative experiences, there is an explanation for the lack of explanation.* In such states, the flow of time stops as one loses his or her grasp on consciousness, and whatever follows while unengaged from the business of living, the language of consciousness inevitably would fail to explain it. There cannot be any memory of being in this vast ocean of ultimate peace and elation. At this stage of self and environmental unawareness (absence of sensing of stimuli), *one becomes nothing and is part of everything and perhaps even knows it all*, maybe because all questions likely lingering in the subconscious, having been addressed, disappear from the scope of the brain. A peaceful eternity of enlightenment prevails, defying any sensible verbalization. Such states—a perception of experiences of unison with nature that cannot be put into words—are often reported by yogis. Anecdotally, similar experiences have also been reported by subjects in controlled LSD experiments (LSD is an extract from consciousness-altering plants), which can also be attributed to momentary mental operation disruptions, which put a halt on normal brain functions along with perhaps some neuronal reconfigurations. The following segment, directly taken from "Neuroscience & Biology; Science and the Wonder of Nature (http://www.eurekalert.org/pub_releases/2016-04/cp-hlc040616. php)" explains the reason for this feeling of "nothingness," described as "ego dissolution" in LSD experimentation:

How LSD can make us lose our sense of self
When people take the psychedelic drug LSD, they sometimes feel as though the boundary that separates them from the rest of the world has dissolved. Now, the first functional magnetic resonance images (fMRI) of people's brains while on LSD help to explain this phenomenon known as "ego dissolution." As researchers report in the Cell Press journal Current Biology on April 13, these images suggest that ego dissolution occurs as regions of the brain involved in higher cognition become heavily over-connected.

As it was explained in the philosopher's stone section, some meditation practices achieve cognition disruption by overwhelming the brain with irresolvable questions. Incidentally, the above report confirms the computational resolution/solution theory put forward in this book.

An anecdotal report from Huffman, the developer of LSD, is noteworthy: he reports the repetition of a childhood dream—a feeling of being in unison with nature—in an inadvertent LSD experiment. Children, who have very little burden of living (consciousness interaction), are naturally more susceptible to such dreams, since the brain is devoid of engagements with unresolved life complexities; in semblance of a meditative state.

The aforementioned ranges of meditative and phenomena-bedazzled states in the minds of early man and mystics are hardly achievable in our present environment for obvious reasons. For one thing, the unavoidable materialistic aspects of living have left an inerasable hum in the brain, unavoidably engaging it and making it hard to suppress. Even more so, the avalanche of wonderful discoveries in the practices of systematic thinking in disciplined research has left very little for discovery in the way of the seemingly endless, prolonged efforts of the old seekers. Nonetheless, the chance, as minute as it may be, for a nonchanneled approach to discovery always exists. The power of deep introspection cannot be discounted, and it is likely that many wonders of our brain's encyclopedia of knowledge are there to be experienced.

The irony of death is that it fully accomplishes the ultimate quietude that the highest meditation strives for—a state of muffled conscious brain activity. It is intriguing to wonder if any events in the brain are perceived at the occurrence of death. It's true that humans relinquish communicable awareness at some point preceding death, as we do in sleep or under anesthesia, but is there any dream-like stage before death? Obviously, there is no precise way of defining the moment of death, before which the brain is still likely to be active. Probably the quiescence of brain waves is a rational criterion. However, the electronics of amplifications have a lot to do with what is seen on the medical screen, the electroencephalogram (EEG). There may well be some period of faint activity in the brain that nobody ever looked at or cared about. If this is true, then it is conceivable that with all the inhibitory forces wiped out, and with total computation capacity available, complex simulations are likely to take place, and perhaps experiences like those in deep meditation occur! The Tibetan "Book of the Dead" discusses practices to prepare one for forty-nine days of journey after death. One should avoid dreams of troubling imagery or adventures until he or she enters in a new higher karmic state, with perhaps pleasant experiences. Such reflections of mythical practices probably originated from the experiences of early mystics at the height of their meditative states. It may be hard to digest, but isn't death an even higher state of the relinquishment of all the material aspects of life? Even more so since all living-related links are also severed as well. The fact that some near-death patients reportedly experience no pain and feel a sense of recovering adds some credence to the latter situation. Given an active state of brain circuitry in the face of no conscious demands on the unconscious during deep meditation, it is not easy to reject the symbolic reports of ancient mystics in their experiences, the hints of possibilities of vast displays of imageries (knowledge) that the unconscious machinery may be holding. The question arises: If a peyote traveler requires days of psyche preparation to achieve a high meditative state, does one not need a proper predisposition, a good karma, to embark on the final journey to experience heaven rather than hell? Do the Tibetans know something?

Means to the End

● ● ●

THE BAFFLING ACCIDENT THAT RENDERED the universe and created life on the dot of our planet, the Big Bang, has left the steering of the runaway truck of survival on a very rough and uncertain terrain, in the control of novice humans who—depending on their brains' whims—have the possibility of learning from the experience of every bump to augment the likelihood of lasting longer while the light persists. However, the road hazards have been overwhelming and confusing, enough to have deprived the majority of humans of much learning and the ability of projection to prepare for what may lie ahead, and hence the chaotic human life persists! As it stands, brains have so far been mostly (*unconsciously*) at the service of individualistic *instinctive* drives of survival, which, in the realm of the human jungle, interprets itself in very selfish behavior by almost all, and for some, in the reckless accumulation of wealth and power, with little or no concern for its perilous impact on the sustainability of the race. History and the devastating state of our living environment speak to this conundrum.

Nonetheless, some among humans, by freaks of nature, gradually began to figure out the dilemma of life. As a result, some progress in the quality of life, and occasional peace and some sense of temporary ease for parts of the human mass, have come about. The evolving findings have enlightened some other for further self-discovery of where their optimum directions are issued from, recognizing ***the role of the brain***.

The human brain, a computational network developed over millions of years, offers fascinating potential for learning, development of peaceful and global harmonious living, and much advancement in revealing more of the workings of universe—of course all within the limits of our perceptions of the realities in themselves. However, human beings generally, with some exceptions, have not been able to take much benefit from the full potential of their brains because of a variety of additional limiting factors. To begin with, they are born carrying certain traits and are raised subject, and bound, to social, geographical, religious, and cultural settings, which, depending on the measure of constraint they exude, tax their brains accordingly throughout life. Humans also commonly suffer from internal and external conflicts and, to varying degrees, from insecurity (e.g., economic, political), which engages them mentally most of the time. Surprisingly, the industrialization and modernization of recent centuries have added more to human mental unease; psychological disorders of various intensities among the masses have become prevalent. Even "normal" people now have comparably more mental struggles and anguishes.

The general absence of a rich mental and cultural upbringing for most humans leaves their minds devoid of the tendency to question their lives, to seek and strive for answers to the complexities affecting them. A major issue that has, and can, at times serve as a barrier to the achievement of harmony and balance in the affairs of humanity at large are religious, ideological, and patriotic indoctrinations, which happen at all levels of societal organizations, from individual families to national levels. Many imprints of the brain, which originate from one's early upbringing (whatever the cause), can easily be exploited for misguided or pernicious intentions by charismatic leaders raising slogans, or by national authorities through artificially created crises, which prompt the imprints to operate as drivers for the intended end. History bears much witness to the atrocities and disasters brought about by such phenomena. Darwin warned us of such designs:

> *It is worthy of remark that a belief constantly*
> *inculcated during the early years of life,*
> *whilst the brain is impressionable, appears*
> *to acquire almost the nature of instinct*
> *and the very essence of an instinct is that*
> *it is followed independently of reason.*

(CHARLES DARWIN, *THE DESCENT OF MAN*, 1871)

The issues with imprints are their nature, and what autonomous actions they can drive. As it has been discussed earlier, and elaborated upon in the works of Mlodinow, much of our behavior is the result of the learning that has been imprinted into the unconscious, Therefore, raising children in environments rich with knowledge and cultures that value harmony among humans and respect for nature renders a well-balanced mind, which by all likelihood will be endowed with enough power of reasoning to resist the temptation of hollow promises and not be sold to slogan and to avoid the urges of senseless pleasures of the ego; such is required for the healthy sustenance of humans and their environment. In this regard, it seems that many other animate beings like ants and bees and perhaps many others in the animal world have come a long way. Apparently they have managed to manipulate the genetic codes of their eggs for certain desired behaviors needed for the sustenance and health of their communities (the recently discovered possibility of gene editing is perhaps putting humanity on such a path, though at this stage of the maturity of human consciousness, it may be a dangerous tool). Whether theirs is a utopia is hard to know, but they do not seem to have coup d'états or torture chambers within their societies, as far as one can tell.

In the context of our computational brain, learning involves programming and reprogramming, which has far-reaching implications in life. Among humans, though, the embedding of the fundamentals of initial behavior and beliefs occurs in the family setting; nonetheless, the most effective types of the mind's programming occur usually in educational systems, the military, prisons, intelligentsia, and religious cults. The requirements for success during exposures are a proper environment and willful young participants, lured in by the appeal of acceptable societal values. It is noteworthy that programming is also sometimes done even on unwilling participants; apparently this requires overcoming the barriers of persona and all its in-place checks and balances, something that allegedly some intelligence organizations occasionally find justifications for doing!

At individual levels, the programmability of the brain allows for the prospect of reevaluating and even altering behavior. The possibility of modification is based on the fact that one is generally born with some measure of nonsocietal compliant behavioral traits—innate individual survival drives going back to the primordial beginning—which are of course susceptible to change in many directions. Teachings of different religious ideologies and psychological and cognitive therapy techniques rely and operate on such premises. Some advanced societies provide the best opportunities for rehabilitation to their societal offenders, even to those who have committed the most heinous crimes.

Again, the roles of the initial seeds of knowing, or wanting to know, for the progression of humanity, cannot be overemphasized.

Most human minds are in general loaded with certain beliefs and may thus be deprived of inquisitiveness about new or alternative lines of thought. Prevalent in some societies is the absence of much societal effort for upgrading the population's knowledge base, which inbreeds resistance to the reevaluation of one's mental stance due to conflict with the earlier established mental dispositions. In general, economic conditions for many people deprive them of any access to unbiased knowledge, or any motivation for acquiring it. Such deprivations cause much malice in such societies; that is why for many, even simple contradictory nuances of thoughts or behavior put them in major conflict with themselves and others in their daily lives. Degrees of absence of conflicts within societies have to do with the resources allocated for education, cultural development, and for the care and welfare of their members. The happiness index among populations in industrial countries is very telling of the effectiveness of such measures!

Religious ethics and morality have helped, at times, in achieving harmonious and less conflicted societal life, at least within the sects—the thousand years of calm and quiet in the Dark Ages (of papacy), despite its many adversities, speaks to this fact. It should not be a surprise that these lines of thoughts, in their essence, speak to the manifests of some religions that subscribe to old teachings as a way to human salvation. Though the role of religion in helping more harmonious societal life cannot be discounted, humans are not yet free from the burden of troubles within themselves. Perhaps the reason lies in the way its edicts place the pleasure-seeking persona in constant inner conflict—and perhaps this is why religion is bending more and more in the way of modernity!

Within the context of behavior modifiability, statements such as "I am this way or that way" or "I have to have this to feel happy" and perhaps hundreds more like them, do not really make sense, especially if they induce suffering and conflict. Since our apparent acceptance or willingness serves as the cornerstone of such behavior that creates mental conflict, its antidote is exposure to tangible data and sustained education and guidance by peers, to fundamentally prove the unhealthiness of such states of mind. Various fields in the humanities are rich with methods and techniques for remedying severe cases of mental unease. Medical fields have also made headway in finding remedies for chemical imbalances that render very difficult behavioral and mental problems. The take-home point here is to integrate these approaches to avoid easy drug-oriented solutions.

Finally, the possibility of a meaningful and livable life for future generations lies in the prevalence of knowledge among the majority of humans of the underlying, unbending principles of sustainability in the face of the delicate balance of nature that dictates the serious limits of exploitation of its resources. For this the human brain needs to be immersed in true knowledge and its resources should be in service to the welfare of all, devoid of any agenda to the detriment of others. True knowledge will lead to new ways of fulfilling our instinctive needs; it will serve in the way of empowering changes that can render not only peace of mind but also awaken the true mystic in us. Avoiding the repetition of what has been discussed in the earlier sections, we draw attention to clues of the power of mind in some ordinary events—like that of the winning gambler we discussed before. We need to stretch our imagination somewhat in order to provide a new perspective on his experience. Let us assume that we have installed a few very fast cameras that are connected to a data acquisition system. Further, we assume that all the intricacies of the Newtonian mechanics of the motion of the roulette ball have been algorithmized in a program and loaded into the same computer that receives the wheel and the ball motion information from the cameras along with the data for the frictional properties involved.

Under such circumstances the computer can conceivably predict the final destination of the ball, i.e., the numbered slot into which it falls. This changes the game of chance to a deterministic problem with a predictable outcome. Pressing the imagination a bit more, one may think that the neuronal net computer of the brain can perhaps do the same calculation. The net gets trained during the gambler's observation period (tuning the web to pose the motion problem correctly), using the input and output for a number of runs of the game until a correct prediction is flashed into the utterance interface! Given such hidden powers within us, humanity is on the right track for deploying them for the betterment of their lives; true progress is inevitable, despite the evils of the misguided survival instinct. This force will eventually be deployed for the true survival of the races and the planet, if we do not destroy all the bridges behind us:

> *If humanity is not cut short, due to self- or nature-caused cataclysms, more of the power of the brain will gradually be deployed to render sustainable conditions for life, and perhaps at some point in the distant future, the all-encompassing treasure trove of man's innate working knowledge of the universe will avail itself as a result of continued efforts at the frontiers of scientific research.*

Final Words

● ● ●

THE ACCEPTED CONCEPT OF THE computational brain was fundamentally furthered in this work through scientific inference as to
its inner workings: computations in the brain are based on its net
structural property of resolving (algorithmization) all exposed phenomena as sets of parametric linear simultaneous equations, and
for much of the biological sustenance of life, the parameters are
known (per specific gene expression at synapses), and instantaneous
solutions of the equations drive the needed autonomous operations.
And for consciousness-related complex phenomena, trial-and-error
(learning) approaches determine the parameters, and solutions
become possible. *As in any complex problem resolution, modeling
of the problem domain where the solution is realized is placental
to the solution; therefore the space-time domain of life is a simulated version (representation) of the real world*—subject to the
limitation of senses—in the brain, which is rendered into conscious
perception.

The fact that many life complexities can be expressed
algorithmically—by sets of simultaneous equations—in the brain
is in the parlance of many scientific and engineering complexities
that get resolved in such formalism before they can be investigated

through computations. Such resolutions of the governing laws of nature, as the ***equations of life and living, in the brain***, continually engage it in solution operations; shaping beings' abilities, defining their brains' states of ***unconsciousness*** and ***consciousness*** and their behavioral dictums along the evolutionary path. Consciousness is the interactive brain display of solutions for beings' entanglement with nature's governing forces, which are updated (for changing conditions) at every waking moment, serving the overall whimsical drive of survival. Self-consciousness, ***"Cogito ergo sum," is the behavioral dictum of survival, driving the higher creatures to do what they perceive to be the results of their thinking or volition.***

The brain operations involved in the development of the consciousness (experience) of the simulated (modeled, represented) world deploy much of the brain's computational power. However, the fact remains that this is only part of the computational tasks of the brain: solving the complexities of daily life and the maintenance of the body's homeostasis are part of its ongoing computational operations. Though beings are equipped with requirements for the natural activities of survival, the mental complexities brought about by modern life may tax the brain to the degree of hampering or disabling its normal operations, either due to an insufficiency of data or the irresolvable nature of the complexities to which we may refer to as stressors. The absence of resolutions results in confusing signals from the brain to the rest of the body, creating secondary adverse effects (unease of various natures), which are attributed to what is vaguely referred to as stress. When homeostasis is in danger, the immune system springs into action to prevent the wrong brain signals, practically disconnecting the signal transmission lines in the process. This causes imminent side effects, neuronal malfunctions called autoimmune disease. However, in the context of the stress-development mechanism in the brain and the proper role of the immune system to prevent adversity, calling the side effect an autoimmune disease is a medical misnomer, since it is actually the result of an immune system performing at its best.

The forces of survival have driven humans toward societal living, and forces of intellect have been steering it toward civilized life. Considering where humans started their collective journey, the gains they have made since their appearance on the planet a short time ago have been impressive, and therefore, one may imagine a peaceful, harmonious, and sustainable existence in the future world. However, much trouble is brewing along the path to the future, and escaping it would require drastic jumps in the collective consciousness of the masses. We need to count on our gray cells for philosophical, ideological, and scientific breakthroughs to make it likely!

Further highlights of this book's contribution can be brought out by drawing attention to the discoveries of the brain's computer potential, the nature of its input/output data, and the identification of its interfaces:

1. The brain, being configured in the evolutionary process, is laden with scripts (programs) for all natural phenomena, and as such it holds an encyclopedic library of knowledge, all embedded as computational patterns (genetically expressed).

2. Scrutinizing the paths of disciplined and undisciplined discoveries, from ancient times to the present, discloses the fact that they happen when the right question is mentally formulated.

3. Vision and thought are perceptions of the brain's computation outputs (efferent signals) at the biological interfaces.

4. Given the successes of the tactile vision substitution system (TVSS) and other methods in enabling vision perceptions

in blind subjects, it leaves no doubt that vision and touch signals are not different in kind but only in intensity, and the former accordingly occupies different and/or bigger brain modalities (network segments) for computations of perception. Vision signals are reflected-matter property-modulated light, which turns into electrochemical signals and passes through the eyes to the brain; metaphorically, beings are touched by nature, contacting millions of nerve endings in the eyes. Bats use reflected-matter (air molecules) modulated sonic signals to see their prey—their remote touch creates vision.

5. Considering the theory of internal language (brain code) and the fact that speech downloads to the vocal cords and that thought can be instantly verbalized, muffled thought (it being also a brain efferent signal) can imperceptibly affect any interface, the likeliest of which is a silent mode of the vocal cords. Astrophysicist Stephen Hawking uses eye movement, which is picked up by EEG, and speech synthesizers to create audio for his presentations.

6. In the context of the computational brain, the reported perception of unison with nature during high meditative states or while under the influence of hallucinogenic drugs allows the proposition of the maxim of consciousness for all matter, inanimate or inanimate, which provides the highly searched-for philosophical axiom.

7. Though the concept of causality has settled the question of free will for many, the computational brain concept closes the chapter on it.

I close with the hope that the fundamental theories introduced in this book—a product of a mental happening—perhaps have resolved the centuries-old philosophical dilemma concerning the nature of thought, vision, free will, consciousness, etc. And last but not least, the autonomous resolution/solution theory of brain computational function could also serve to draw attention to the need for improved jurisprudence, social justice, and finally human societal ideologies.

BIBLIOGRAPHY

Achiron, A., et al. 2012. "Multiple Sclerosis in Israeli Children: Incidence and Clinical Cerebrospinal Fluid and Magnetic Resonance Imaging Findings." *IMAJ*: 14.

Aghajan, Z. M., L. Acharya, J. J. Moore, J. D. Cushman, and C. Vuong, et al. 2015. "Impaired Spatial Selectivity and Intact Phase Precession in Two-Dimensional Virtual Reality." *Nature Neurosciences* 18: 121–125.

Alaert, K., S. P. Swinnen, and N. Wendroth. 2009. "Is Human Primary Motor Cortex Activated by Muscular or Direction-Dependent Features of Observed Movements?" *Cortex* 45: 1145–1155.

Arbib, A. M. 1989. *The Metaphorical Brain 2, Neural Network and Beyond*. New York: John Wiley & Sons.

Avey E. A. 1954. *Handbook in the History of Philosophy*. New York, Barnes and Noble

Bach-y-Rita, P., C. C. Collins, A. F. Sanders, B. White, and L. Scadden. 1968. "Vision Substitution by Tactile Image Projection." *Nature* 221: 963–964.

Bach-y-Rita, P. 2006. "Tactile Substitution Studies." *Annals of New York Academy of Sciences* 1013: 83–91.

Bach-y-Rita, P., K. A. Kaczmarek, M. E. Tyler, and J. Garcia-Lara, J. 1998. "Form Perception with a 49-Point Electro-Tactile Stimulus Array on the Tongue: A Technical Note." *J Rehabil, Res Dev.* 35 (4): 427–30

Barkley, R. J. 1939. "An Informal Exposition of Proofs of Gödel's Theorem and Church's Theorem," reprinted from the *Journal of Symbolic Logic* 4; 53–60, in *The Undecidable* by Martin Davis (loc. cit.), 1965, 223–230.

Baum, E. 2004. *What Is Thought?* A Bradford Book, Massachusetts: MIT Press.

Beck, S. J., and J. B. Molish. 1959. *Reflexes to Intelligence.* Free Press.

Bostrom, N. 2003. "Are You Living In a Computer Simulation?" *Philosophical Quarterly* 53 (211): 243–255.

Broad, C. D. 1978. *Kant, an Introduction.* MA: Cambridge University Press.

Campbell, J., and B. Moyers. 1988. <u>*The Power of Myth,*</u> edited by <u>Betty Sue Flowers</u>. Doubleday.

<u>Center for Cognitive Ubiquitous Computing</u> at the <u>Arizona State University</u>. 2016. https://cubic.asu.edu/

Chalmers, D. C., D. C. Dennet, and R. Hufstadter, et al. 2009. *Mind and Consciousness: 5 Questions.* Vince Inc; Copenhagen.

Charcot, J. M. 1877. *Lectures on the Diseases of the Nervous System, Delivered at La Salpêtrière.* London: New Sydenham Society

Chomsky, N. 2007. *On Language.* New York: New York Press.

Churchland, P. S., and R. Grush. 1999. "Computation and the Brain." MIT Encyclopedia of Cognitive Sciences.

Churchland, P. S. and T. J. Seinowski. 1992. *The Computational Brain*. Cambridge, MA: MIT Press.

Colman, B., and J. Moyne. 2004. *The Essential Rumi*. Harper.

DAEDALUS. 1988. "Artificial Intelligence." *Journal of the American Academy of Arts and Sciences*.

Dantzer, R., and W. K. Kelley. 1989. "Stress and Immunity: An Integrated View of Relationships between the Brain and the Immune System." *Life Sciences* 44; 26: 1995–2008.

Dasen, P. 1994. "Culture and Cognitive Development from a Piagetian Perspective." In *Psychology and Culture*, edited by W. J. Lonner and R. S. Malpass. Boston: Allyn and Bacon.

De Kloet, E. Ron, and F. Holsboers. 2005. "Stress and the Brain from Adaptation to Disease." *Nature Reviews: Neuroscience* 23 (6): 463–475f.

Durant, W. 1991. *The Story of Philosophy*. Simon & Schuster.

Edelman, G. 1987. *Neural Darwinism. The Theory of Neuronal Group Selection*. New York: Basic Books.

Freud, S. 1935. *A General Introduction to Psychoanalysis*. New York: Boni and Liveright.

Keysers, C., B. Wicker, V. Gazzola, J. L. Anton, L. Fogassi, and V. Gallese. 2004. "Touching Sight SII/PV Activation during the Observation and Experience of Touch." *Neuron* 42 (2): 335–346.

Keysers, C., and V. Gazzola. 2009. "Expanding the Mirror: Vicarious Activity for Actions, Emotions and Sensations. *Current Opinion in Neurobiology* 19: 666–671.

Schrodinger E. 1944. *What Is Life?* Cambridge: Cambridge Press.

YouTube. 2011. "Helen Keller Meeting President Eisenhower and Going Shopping." http://www.youtube.com/watch?v=OL3ITXHOTo8.

Gail, F. 1999. *Plato 1: Metaphysics and Epistemology.* Oxford, England: Oxford university press.

Geldard, F. A., and C. E. Sherrick. 1965. "Multiple Cutaneous Stimulations; the Discrimination of Vibratory Pathforms." *Journal of the Acoustical Society of America*: 37: 797–801.

Gödel, K. 1931. *On Formally Undecided Propositions.* New York: Basic Books.

Hawking, S. 2010. *The Grand Design.* New York: Bantam Books.

Hofstadter D. and D. C. Dennett. 1981. *The Mind's I.* Bantam Books.

Hufstadter, Douglas R. 1979. *Gödel, Escher, Bach: An Eternal Golden Braid.* Vintage Books.

Humphreys, L. G. 1979. "The Construct of General Intelligence." *Intelligence* 3 (2): 105–120.

Izad, S., et al. 2013. "Significant Increase in the Prevalence of Multiple Sclerosis in Iran: A Report on the Patients Receiving Beta-Interferon in Different Provinces." *Ir. J Neurology* 12.

Jung, C. G. 1933. *Modern Man in Search of a Soul.* New York: Harcout. Inc

Juyang, W. 2012. *Natural and Artificial Brain, Introduction to Computational Brain.* BMI Press.

Janglova, D. 2004. "Neural Networks in Mobile Robot Motion." *International Journal of Advanced Robotic Systems* 1 (1).

Kandel, R. C., H. J. Schwartz, and T. M. Jessell. 2000. *Principles of Neurosciences.* New York: McGraw-Hill.

Kandel, Eric. 2013. *The New York Times*, Sunday Review, Opinion Page, 12/1.

Kerr, N., and G. W. Donhoff. 2004. "Do Blinds Literally 'See' in Their Dreams?" *Dreaming* 14: 23–233.

Kounios, J., and M. Beeman. 2015. *The Eureka Factor: Aha Moment, Creative Insights and Brain Science.* New York: Random House.

LSD experiment. RadioLab. http://www.radiolab.org/story/painting-tongues/.

Mlodinow, L. 2011. *Subliminal: How Your Unconscious Minds Rules Your Behavior.* New York: Vintage Books

McCulloch, S. W. and W. H. Pitts 1993. A logical calculus of ideas immanent in nervous activity. *Bulletin of Mathematical biophysics*; 5; 115-133.

Mohr, C. David. 2012. "Stress Management Prevents Brain Lesions in Multiple Sclerosis." *Northwestern University Report.*

Myers, G. D. 2000. *Psychology.* New York: Worth.

National MS Society. 2009. "History of MS." www. Nationalmssociety. org.

Ortiz, T., J. Poch, J. M. Santos, C. Requena, and A. M. Martinez, et al. 2011. "Recruitment of Occipital Cortex during Sensory Substitution Training Linked to Subjective Experience of Seeing in People with Blindness." *PLOS ONE Journal* 6 (8): 1371.

Penrose, R. 1990. *Emperor's New Mind.* Oxford, England.

Prabhupada, S. 1972. *Bhagvad-Gita As It Is: HDG A.C.* Bhaktivedanta Book Trust.

Rinpoche, Sogyal. 1994. *The Tibetan Book of Living and Dying.* San Francisco: Harper Collins.

Russell, B. 1910. *Logic and Knowledge Essays 1950.* New York: Cambridge Books.

Sapolsky, Robert, Lewis C. Krey, and Bruce S. McEwen. 2000. "The Neuroendocrinology of Stress and Aging: The Glucocorticoid Cascade Hypothesis." *Science of Aging Knowledge Environment* 38: 21.

Sapolsky, Robert. 1990. "Stress in the Wild." *Scientific American* 262 (1): 106–113.

Searle, J. 1999. *An Essay on Philosophy of the Mind.* Cambridge: University Press.

Schwartz, S. 2009. *Visual Perception: A Clinical Orientation.* New York: McGraw Hill Medical.

Schad J. N. 016. "Brain Neurological Constructs: The Neuronal Computational Schemes for Resolution of Life Complexities." *Journal of Neurology and Neurophysiology;* 4: 5.

Schad, J. 2016. "Neurological Nature of Vision and Thought, and Mechanisms of Perception Experiences. *Neurol Stroke. 4:5*

Schad, J. 2015. "Mental Stress: Brain's Complexity Resolution Dilemma. *Neurol Stroke. 2:4*

Schad, J. 2013. "Stress Caused Adverse Entanglement of the Nervous and Immune Systems: A Case for MS." *Medical Hypothesis* 80: 156–157.

Searle, J. 1989. "Artificial Intelligence and the Chinese Room: An Exchange." *New York Review of Books* 36: 2.

Skinner, B. F. 1984. "The Operational Analysis of Psychological Terms." *Behavioral and Brain Sciences* 7 (4).

Sternberg, R. J., and W. Salter. 1982. *Handbook of Human Intelligence.* Cambridge, UK: Cambridge University Press.

Selye, H. 1955. "Stress and Disease." *Science* 122: 625–631.

Stace, W. T. 1973. *Mysticism and Philosophy*. London: Macmillan.

Turkle, S. 1988. "Artificial Intelligence and Psychoanalysis: A New Alliance." *DEDALUS*, Winter Edition.

Turing, A. 1948. "Machine Intelligence," in Copeland, B. Jack, *The Essential Turing: The ideas that gave birth to the computer age.* Oxford: Oxford University Press.

WHO. 2008. *MS Atlas.*

Wechsler, D. 1944. *The Measurement of Adult Intelligence*. Baltimore: Williams & Wilkins.

Wienpahl, P. 1964. *The Matter of ZEN*. New York University Press.

Wittgenstein L. Last Writing on Philosophy of Psychology. 1982. Chicago: Chicago University Press.

www.ingramcontent.com/pod-product-compliance
Lightning Source LLC
Chambersburg PA
CBHW070320190526
45169CB00005B/1679